SpringerBriefs in Computer Science

Series Editors

Stan Zdonik
Peng Ning
Shashi Shekhar
Jonathan Katz
Xindong Wu
Lakhmi C. Jain
David Padua
Xuemin Shen
Borko Furht
V. S. Subrahmanian
Martial Hebert
Katsushi Ikeuchi
Bruno Siciliano

T0183878

For further volumes:
http://www.springer.com/series/10028

Hongzi Zhu · Minglu Li

Studies on Urban Vehicular Ad-hoc Networks

 Springer

Hongzi Zhu
Minglu Li
Department of Computer Science and Engineering
Shanghai Jiao Tong University
Shanghai
People's Republic of China

ISSN 2191-5768   ISSN 2191-5776   (electronic)
ISBN 978-1-4614-8047-1   ISBN 978-1-4614-8048-8   (eBook)
DOI 10.1007/978-1-4614-8048-8
Springer New York Heidelberg Dordrecht London

Library of Congress Control Number: 2013940443

Printed on acid-free paper

Springer is part of Springer Science+Business Media (www.springer.com)

Preface

Vehicular Ad hoc Networks (VANETs) are emerging as a new technology to provide a wide spectrum of safety, efficiency, and comfort applications to the public and governments. It has been immensely successful and naturally attracted considerable attention from both academia and industry since its introduction about one decade ago. Numerous publications and projects have been devoted to this topic. However, the realistic behavior of the network at a large scale is still unavailable today, due to the initial stage of VANET deployments. The goal of this book is to offer some fundamental observations of node and network behavior when the network scale reaches over 10,000 vehicles and demonstrates mobile sensing applications based on VANETs in urban scenarios. The target audiences are researchers interested in getting to know VANETs, in particular graduate students. It is also our hope that this book can be useful to experts as quick reference.

This book starts with an introduction on VANETs and representative experimental work in the world such as the MIT CarTel project, the UMASS DieselNet project, and the GM DSRC Fleet. We then introduce the empirical studies conducted in SJTU, which are based on three realistic GPS data sets collected from taxies and buses in Shanghai and Shenzhen, two metropolises in China.

In Chap. 2, we describe the characteristics of the trace data and main challenges and issues in data analysis. In Chap. 3, we extensively study the distribution of inter-contact time (ICT) between a pair of vehicles and establish a general vehicular mobility model in urban settings which follows the observed ICT distribution. Some of the proofs are involved and can be safely skipped at first reading. Nevertheless, we decided to include them because they either illustrate useful analytical skills or provide details that are missing in the original papers. Due to the limited time, space, and of course our knowledge and ability, the content of this book is far from extensive.

Chapter 4 covers two opportunistic data forwarding strategies in VANETs. In this chapter, first, the temporal correlations between pairwise contacts are analyzed and further utilized to predict future contact information between vehicles. As data are relayed in VANETs in a store-carry-forward fashion, such estimated future contact information can be leveraged to improve routing performance. Moreover, the sociality of vehicular networks is also examined and we have the observation that vehicles do have clear social relationships that can further stimulate the routing performance. We describe two proposed opportunistic routing schemes in VANETs which utilize such knowledge and gain better performance in terms of end-to-end delay and network traffic cost.

Chapter 5 introduces a distributed online vehicle tracking scheme in large cities. RFID systems are deployed to capture vehicles and location information about vehicles is locally stored among a large number of nodes distributed in the city. The main challenge is to guarantee that the response time of a query issued from anywhere in the city meets a given real-time requirement and meanwhile to minimize the network cost for location updating and query forwarding in the network. We describe a scheme which organizes the nodes into different regions. With this organization, location information updating is restricted within a small scale and still keeps the whole network updated. In addition, the query can be forwarded to the most up-to-date node within the given time requirement.

Chapter 6 covers a mobile sensing application which uses commuting vehicles as mobile sensors to sample the traffic condition on surface roads and analyzes these sensory data to infer the traffic condition on those roads with insufficient sample data. The main challenge is to remove noise embedded in the data and recover the traffic condition information from lossy sensory data.

We would like to express our greatest appreciation to Prof. Xuemin (Sherman) Shen for providing the opportunity to write this brief book for Springer. Especially, Hongzi is greatly indebted to Prof. Lionel Ni for introducing him to the field and guiding him in his research. Hongzi also owes deep gratitude to his post-doctoral supervisor Prof. Xuemin (Sherman) Shen for his continuous support and guidance. Hongzi would like to acknowledge his wife, Dr. Shan Chang, who not only provided valuable comments on the writing of the book but also encouraged him throughout the process. We are grateful to all our collaborators and colleagues, in particular, Dr. Yanmin Zhu, Dr. Guangtao Xue, Dr. Xinbin Wang, and his Ph.D. student Luoyi Fu, who made great contribution in our published papers and this book. We also would like to thank Springer, especially Ms. Melissa Fearon and Ms. Courtney Clark, for their support in various aspects in the editing and publishing of this book.

Shanghai, People's Republic of China,   Hongzi Zhu
May 1, 2013   Minglu Li

Contents

Chapter 1
Overview

1.1 Introduction to Vehicular Ad-hoc Network

Vehicular Ad hoc Networks (VANETs) are emerging as a new landscape of mobile ad hoc networks, aiming to provide a wide spectrum of safety and comfort applications to drivers and passengers. In VANETs as illustrated in Fig. 1.1, vehicles equipped with wireless communication devices can transfer data with each other (*inter-vehicle* or *V2V communications*) as well as with the roadside infrastructure (*vehicle-to-roadside* or *V2I communications*). Combined with various sensors, such as image/xxx sensor, accelerometer, GPS receiver and radar, and an embedded processing unit, vehicles appear "smarter" than ever, having a better understanding about the surrounding environment and other vehicles on the move. Both the new sensing and wireless communication technologies enable the promising applications of VANET in the future with respect to safety, efficiency of infrastructure and comfort. Foreseeing this trend, both academia and industry put great efforts in investigating the new possibilities that can be brought by VANETs. During the past two decades, a vast number of projects and institutes related to VANET have sprung up, trying to study research problems as follows:

- **Short range wireless communication technology:** focuses on providing fast wireless links for both V2V and V2I communications in VANETs, devised to work on a dedicated spectrum. For example, in October 1999, the United States Federal Communications Commission (FCC) allocated in the USA 75 MHz of spectrum in the 5.9 GHz band for Dedicated Short-Range Communications (DSRC) [1] to be used by Intelligent Transportation Systems (ITS). In August 2008, the European Telecommunications Standards Institute (ETSI) also allocated 30 MHz of spectrum in the 5.9 GHz band for ITS [2]. The reason of using the spectrum in the 5 GHz range is due to its spectral environment and propagation characteristics, which are suited for vehicular environments. Specifically, waves propagating in this spectrum can offer high data rate communications for pretty long distance (up to 1000 m) with low weather dependence. With the dedicated spectrums, new MAC protocols operating on

H. Zhu and M. Li, *Studies on Urban Vehicular Ad-hoc Networks*,
SpringerBriefs in Computer Science, DOI: 10.1007/978-1-4614-8048-8_1,
© The Author(s) 2013

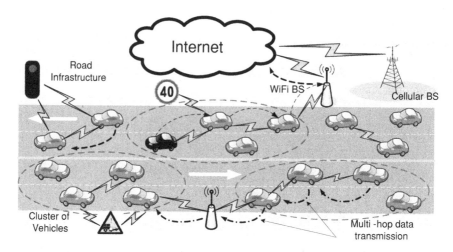

Fig. 1.1 An illustration of VANET, where vehicles can "talk" to roadside infrastructure such as WiFi access points, traffic lights and speed limitation signs via V2I communications and to other vehicles via V2V communications. In the figure, *dashed arrow lines* represent the data transmission paths

these spectrums are designed in order to provide fast and robust links between mobile devices. How such MAC protocols perform in complicated vehicular environments should be carefully studied and verified before it can be applied into real applications.

- **Mobility model analysis:** studies how vehicles move in the network. As the entities in VANETs are highly mobile vehicles, the fundamental characteristics of vehicular mobility, such as how vehicles rendezvous in terms of frequency and duration, how they visit a location and how wide they can cover a region of interest in both space and time dimensions, are therefore crucial to the design and ultimate performance of network protocols. In the literature, most studies focus on theoretical models, such as random walk, random way point. While theoretical mobility models facilitate problem analysis, they are far beyond reality and not practical in designing networking protocols for real systems and their performance analysis. Realistic vehicular mobility model analysis has therefore become a recent hot research area.

- **Opportunistic DTN routing:** considers only V2V communications to forward data between mobile vehicles with the goal of eventually reaching a destination. As two vehicles need to geographically "meet" (i.e., within each other's communication range) before any data exchange, data transfer, therefore, arises in a *store-carry-forward* fashion, which results in long end-to-end delay as in Delay-Tolerant Networks (DTNs). Because establishing an optimal forwarding path in advance between the source and destination in VANETs is very hard even if all future movement of vehicles are known, to design an efficient opportunistic routing algorithm in VANETs is a hard problem to solve.

- **Mobile sensing applications:** aim to leverage the mobility of vehicles to collect environmental information in a large area of field with only moderate cost for system deployment compared with statically installed sensor networks. Due to individual reading errors and sparse sensing data distribution, to gain an accurate map of measurements in the field is very challenging. By advanced data processing techniques such as data fusion among neighboring vehicles, it is possible to know the true calibration of some sensors in the system and make accurate estimates in locations even without any sensor readings.
- **Intelligent Transportation System applications:** aim to improve transport outcomes such as transport safety, congestion control, travel reliability, informed travel choices, environmental performance and network operation resilience. Intelligent Transportation System (ITS) is not a new concept, which have been studied since 1960s and widely implemented in the developed world especially in the United States, Europe and Japan. Comparing with traditional ITS implementation, the new information and communication technology such as sensing technology and VANET present a set of relatively low-cost methods for obtaining travel information along streets, highways, freeways, and other transportation routes. In addition, new ITS applications such as active and coordinative safety between vehicles continuously emerge which are enabled with the new technology.

In the next section, we will introduce the most representative experimental work on VANETs worldwide. For each work, we first describe its background and research goals. We will introduce empirical studies on urban VANETs using real trace data in Shanghai in the following chapters.

1.2 Representative Experimental Work Worldwide

1.2.1 MIT CarTel

The CarTel project at Massachusetts Institute of Technology (MIT) [3, 4] combines mobile computing and sensing, wireless networking, and data-intensive algorithms running on servers in the cloud to address the grand challenges to the efficiency and the safety of road transportation. CarTel is a distributed, mobile sensing and computing system using phones and custom-built on-board telematics devices, which might be thought of as a "vehicular cyber-physical system". CarTel's research contributions include traffic mitigation, road surface monitoring and hazard detection (the Pothole Patrol), vehicular networking, privacy protocols, intermittently connected databases, and the design of multiple generations of in-car hardware using only WiFi for connectivity. In this book, we put emphasis on introducing the vehicular networking part in CarTel.

1.2.1.1 Testbed Setup

In CarTel, 27 cars with custom-made on-board devices form a running testbed, upon which all software and applications are deployed. A typical on-board device consists of a small yet powerful embedded computer, a commodity GPS unit, a miniPCI WiFi card, and other sensors such as 3D accelerometer and camera. The embedded computer has a 586-class processor running at 266 MHz with 128 MB of RAM and 1 GB (or more) of Flash, running Linux 2.6. The GPS unit is connected the computer via USB interface. In addition, an OBD-to-serial adapter is used to allow the embedded computer to access the internal computer of a car made after 1996. Figure 1.2 illustrates the original and upgraded versions of the implementation.

1.2.1.2 Research and Experiments

- *Cabernet:* is a system for delivering data to and from moving vehicles using open 802.11 (WiFi) access points encountered opportunistically during travel. Using open WiFi access from the road can be challenging. Network connectivity in Cabernet is both fleeting (access points are typically within range for a few seconds) and intermittent (because the access points do not provide continuous coverage), and suffers from high packet loss rates over the wireless channel. On the positive side, WiFi data transfers, when available, can occur at broadband speeds. In Cabernet, two new components [5] were proposed for improving open WiFi data delivery to moving vehicles: The first, QuickWiFi, is a streamlined client-side process to establish end-to-end connectivity, reducing mean connection time to less than 400 ms, from over 10 s when using standard wireless networking software. The second part, CTP, is a transport protocol that distinguishes congestion on the wired portion of the path from losses over the wireless link, resulting in a 2x throughput improvement over TCP. To characterize the amount of open WiFi capacity available to vehicular users, Cabernet

Fig. 1.2 The original (*left*) and upgraded (*right*) versions of the implementation of MIT on-board unit

was deployed on a fleet of ten taxis in the Boston area. The long-term average transfer rate achieved was approximately 38 MB/h per car (86 kbit/s), making Cabernet a viable system for a number of non-interactive applications.

- *CafNet (carry and forward network):* is a delay-tolerant stack that enables mobile data mulling and allows data to be sent across an intermittently connected network [6]. CafNet delivers data between nodes even when there is no synchronously connected network path between them. For example, these protocols could be used to deliver data from sensor networks deployed in the field to Internet servers without requiring anything other than short-range radio connectivity on the sensors (or at the sensor gateway node). Different from traditional automotive telematics systems that rely on cellular or satellite connectivity, the CarTel embedded in-car device (i.e., when data is collected using the OBD-connected hardware) should use wireless networks opportunistically. It uses a combination of WiFi, Bluetooth, and cellular connectivity, using whatever mode is available and working well at any time, but shields applications from the underlying details. Applications running on the mobile nodes and the server use a simple API to communicate with each other. CarTel's communication protocols handle the variable and intermittent network connectivity.
- *Wi-Fi Monitoring:* is to map the proliferation of 802.11 access points in the Boston metro area [7]. In this task, a measurement study carried out over 290 "drive hours" over a few cars under typical driving conditions, in and around the Boston metropolitan area. With a simple caching optimization to speed-up IP address acquisition, it was found that for the experimental driving patterns the median duration of link layer connectivity at vehicular speeds is 13 s, the median connection upload bandwidth is 30 Kb/s, and that the mean duration between successful associations to APs is 75 s. It was also found that connections were equally probable across a range of urban speeds (up to 60 km/h). The end-to-end TCP upload experiments had a median throughput of about 30 KB/s, which is consistent with typical uplink speeds of home broadband links in the US. The median TCP connection is capable of uploading about 216 KB of data. The conclusion is that grassroots Wi-Fi networks are viable for a variety of applications, particularly ones that can tolerate intermittent connectivity.

1.2.2 UMass DieselNet

DieselNet [8] is a bus-based DTN testbed that was built from 2004 at University of Massachusetts (UMass), Amherst, USA. The DieselNet operates daily from the UMass Amherst campus and covers the surrounding county. Now DieselNet is part of UMass GENI testbed, and it is open for public experiments.

1.2.2.1 Testbed Setup

DieselNet currently consists of 35 buses each with a Diesel Brick, which is based on a HaCom Open Brick computer (P6-compatible 577 MHz CPU, 256 MB RAM, 40 GB hard drive, Linux OS). Figure. 1.3 shows a typical hardware configuration deployed on a DieselNet bus. The brick is connected to three radios: an 802.11b Access Point (AP) to provide DHCP access to passengers and passersby, a second USB-based 802.11b interface that constantly scans the surrounding area for DHCP offers and other buses, and a longer-range MaxStream XTend 900 MHz radio to connect to road-side device, called "throwboxes". Additionally, a GPS device records times and locations. The custom software allows researchers to push out application updates, take mobility, AP-to-bus connectivity, and bus-to-bus throughput traces. Besides the embedded computers deployed on buses, in DieselNet, stationary and battery-powered nodes with storage and processing are also installed at road side to enhance the capacity of DTNs. Figure 1.4 illustrates the internals of the throwbox prototype.

1.2.2.2 Research and Experiments

- *DTN routing:* Routing protocols for disruption-tolerant networks (DTNs) use a variety of mechanisms, including discovering the meeting probabilities among nodes, packet replication, and network coding. Implemented on DieselNet, MaxProp [9], a protocol for effective routing of DTN messages, is based on prioritizing both the schedule of packets transmitted to other peers and the schedule of packets to be dropped. These priorities are based on the path likelihoods to peers according to historical data and also on several complementary mechanisms, including acknowledgments, a head-start for new packets, and lists of previous intermediaries. In contrast, RAPID [10], an *"intentional"* DTN routing protocol, was proposed that can optimize a specific routing metric such as the worst-case delivery delay or the fraction of packets that are delivered within a deadline. Specifically, in RAPID protocol, the DTN routing problem is

Fig. 1.3 A typical hardware configuration deployed on a DieselNet bus: an embedded computer, 802.11b AP, 802.11b card, and GPS

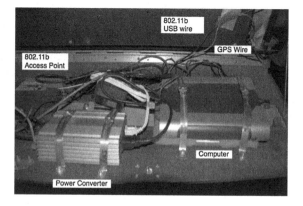

Fig. 1.4 The internals of the throwbox prototype

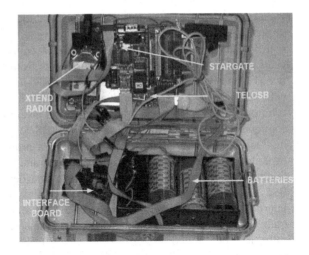

formulated as a resource allocation problem, where resources are allocated to packets to optimize an administrator-specified routing metric. At each transfer opportunity, a RAPID node replicates or allocates bandwidth resource to a set of packets in its buffer, in order to optimize the given routing metric. Packets are delivered through opportunistic replication, until a copy reaches the destination. As DTNs are resource constrained networks in terms of transfer bandwidth, energy, and storage, RAPID makes the allocation decision by first translating the routing metric to a per-packet utility and the first packet to be replicated is the one that provides the highest increase in utility per unit resource used. In addition, to have a local view of the global network state, an in-band control channel is used to exchange network state information among nodes.

- *Network capacity enhancement:* In VANET, data transmission relies on intermittent contacts between mobile nodes using a store-carry-forward paradigm. To enhance the capacity of the network, dedicated road-side "throwboxes" are utilized to increase the opportunities and efficiency of vehicular contacts in DieselNet [11,12]. The hardware of a throwbox uses a multi-tiered, multi-radio, scalable, solar powered platform. The throwbox employs an approximate heuristic for solving the NP-Hard problem of meeting an average power constraint while maximizing the number of bytes forwarded by the throwbox. In DieselNet, the effect of different types of infrastructure, e.g., disconnected relays, base stations connected to a wired backbone network, and wireless mesh network, to the performance of VANET is thoroughly studied [13]. Two key observations were found: First, if the average packet delivery delay in a vehicular deployment can be reduced by a factor of two by adding n base stations, the same reduction requires $2n$ mesh nodes or $5n$ relays. Given the high cost of deploying base stations, relays or mesh nodes can be a more cost-effective enhancement; second, it was observed that adding small amount of infrastructure is vastly superior to even a large number of mobile nodes capable of routing to one another, obviating the need for mobile-to-mobile disruption tolerant routing schemes.

- *WiFi connectivity:* To investigate whether the ubiquity of WiFi can be leveraged to provide cheap connectivity from moving vehicles for common applications such as Web browsing and VoIP, a study of connection quality available to vehicular WiFi clients based on measurements from DieselNet was conducted. It was found that current WiFi handoff methods, in which clients communicate with one base station at a time, lead to frequent disruptions in connectivity. In addition, it was also found that clients can overcome many disruptions by communicating with multiple base stations (BSes) simultaneously. These findings lead to the development of *ViFi* [14], a protocol that opportunistically exploits BS diversity to minimize disruptions and support interactive applications for mobile clients. In ViFi, a vehicle first designates one of the nearby BSes as the anchor, who is responsible for the vehicle's connection to the Internet. It also designates other nearby BSes as auxiliary, who help to relay traffic in the communication between the vehicle and the anchor BS. Specifically, in order to notify nearby BSes which BSes have been chosen to serve either as the anchor or auxiliary BSes, the vehicle embeds the identity of the current anchor and auxiliary BSes in the beacons that it broadcasts periodically. When the vehicle transmits a packet p to the anchor, if the anchor receives p, it broadcasts an ACK. If an auxiliary overhears p, but within a small time window has not heard an ACK sent from the anchor, it probabilistically relays p. If the anchor receives the relayed p and has not already sent an ACK, it broadcasts an ACK. If the vehicle does not receive an ACK within a retransmission interval, it retransmits p. In this way, the disruptions of Internet access can be minimized. Verified through trace-driven simulations, ViFi doubles the number of successful short TCP transfers and doubles the length of disruption-free VoIP sessions compared to an existing WiFi-style handoff protocol.
- *Mobility model study:* To study the performance of routing protocols and applications in VANET, it is of great importance to accurately characterize transfer opportunities between vehicles. Based on the traces taken from DieselNet, contacts between buses were recorded as they travel their routes [15]. It was found that the all-bus-pairs aggregated inter-contact times show no discernible pattern. However, the inter-contact times aggregated at a route level exhibit periodic behavior. Based on analysis of the deterministic inter-meeting times for bus pairs running on route pairs and consideration of the variability in bus movement and the random failures to establish connections, route-level models were constructed to capture the above behavior.

1.2.3 GM DSRC Fleet

The emergent IEEE 802.11 p-based Dedicated Short Range Communication (DSRC) standard is one of the IEEE 802.11 standards customized for highly mobile, severe-fading vehicular environments. DSRC-based Vehicle Safety

Communications (VSC) systems have attracted great attention from the automotive industry and government agencies because of their simplicity and low cost. As a pioneer, the automotive company General Motor (GM) developed a fleet of three vehicles, on which a vehicular communication system was mount.

1.2.3.1 Testbed Setup

This system consists of four components [16]: (1) DSRC-compatible Radio: Such a radio is built upon the Atheros AR5000 chipset. The default values for transmission power and data rate are 20 dBm and 6 Mbps, respectively. The radios operate in the IEEE 802.11 p 'Wave BSS (WBSS)' mode. The RSSI sensitivity level of successfully received packets is up to −94 dBm. The omni-directional antenna connected to the DSRC radio is mounted on the vehicle roof. The gain of antennas used in our systems is 0 dB; cables and connectors introduce 2 dB signal attenuation. (2) GPS Receiver: The GPS receiver synchronizes to the clock of satellites at a rate of 5 Hz. (3) DSRC Protocol Stack: The prototype system on each vehicle sends out broadcast packets via its DSRC radio every 0.1 s. Each packet is tagged with a vehicle ID and a unique packet sequence number. (4) Vehicle Safety Communications (VSC) Applications: These VSC applications include Stop or Slow Vehicle Advisor (VSA), Emergency Electronic Brake Light (EEBL), Lane Change Advisor (LCA), and Cooperative Collision Warning (CCW).

1.2.3.2 Research and Experiments

- *DSRC measurements:* In the experiment [16], a large volume of experimental data was collected via a series of measurement campaigns using a fleet of three vehicles equipped with the prototype systems. These measurements were conducted in the Detroit metropolitan area, Michigan, from July 2005 to Sep 2007. Five typical environments were considered: (1) urban freeway: eight-lane freeway with a large number of walls, tunnels and overhead bridges, as well as heavy vehicle traffic; (2) rural freeway: six-lane freeway with open lands, less traffic than urban freeway; (3) rural road: two-lane street with heavy traffic; (4) suburban road: six-lane suburban streets with light traffic; (5) open field: no buildings and other vehicles. Through the experiment, several key observations were found: first, the reliability of DSRC presents dominating Gray-zone behavior (i.e., intermediate loss rate); second, the propagation environment has major impact on DSCR characteristics; third, Doppler effect does not seem to significantly impact DSRC characteristics; fourth, reduced transmission power only generates minor degradation in DSRC reliability, which suggests a smaller power (i.e., 15 dBm) rather than default value (20 dBm); fifth, default value of 6 Mbps is a reasonable data rate parameter; and last, both temporal and spatial correlation of DSRC performance are weak in vehicular environments.

1.2.4 Germany FleetNet Project

The project "FleetNet—Internet on the Road" (2000–2003) [17] was set up by a consortium of six companies and three universities: DaimlerChrysler AG, Fraunhofer Institut für offene Kommunikationssysteme (FOKUS), NEC Europe Ltd., Robert Bosch GmbH, Siemens AG, TEMIC Speech Dialog Systems GmbH, Universities of Hannover and Mannheim, and Technische Universität Hamburg-Harburg and Braunschweig. The main objective of FleetNet was to develop and demonstrate a platform for inter-vehicle communication systems. Appropriate applications for demonstration were implemented to show the benefit of inter-vehicle communication systems. A study on business cases and market introduction strategies complemented the technical objectives and the project results were opened to appropriate international standardization bodies.

1.2.4.1 Testbed Setup

Ten Smart cars and a number of roadside stations act as a "real world" testbed. These experimental vehicles are equipped with cabin mounted cameras, LCD touch screens, and internal computers providing access to the car's navigation system and to its body electronics via a CAN bus interface.

1.2.4.2 Research and Experiments

- *Routing and forwarding strategies:* A forwarding method called "contention-based forwarding" (CBF) [18–21] was designed, where the next hop in the forwarding process is selected through a distributed contention process based on the exact current positions of all neighbors. Similar with the medium access control in local area networks such as WiFi, a timer is set for each neighboring vehicle to contend the opportunity to forward a packet. Instead of randomly selecting a timer, the time of a neighboring vehicle is set to a short value if the corresponding vehicle is close to the destination of a packet. In this way, the closer a neighboring vehicle is to the destination, the higher probability it will win the contention for relaying the packet. Together with DaimlerChrysler AG, a position-based router was implemented for inter-vehicle communications. To evaluate the design of the router, a test network of 6 DaimlerChrysler Smart cars was set up. All experimental cars are equipped with GPS receivers, IEEE 802.11 WLAN NICS with planar antennae and the custom router. The test network allows global monitoring of the ad hoc network via GPRS. Performance evaluation of position-based routing with respect to vehicular networks was conducted in both highway and city scenarios.

1.2.5 Europe Network on Wheels (NOW) Project

NOW [22] is a German research project which is supported by Federal Ministry of Education and Research, founded by Daimler AG, BMW AG, Volkswagen AG, Fraunhofer Institute for Open Communication Systems, NEC Deutschland GmbH and Siemens AG in 2004. Besides the partners the Universities of Mannheim, Karlsruhe and Munich and the Carmeq GmbH co-operate within NOW. The main objectives are to solve technical key questions on the communication protocols and data security for car-to-car communications and to submit the results to the standardization activities of the Car2Car Communication Consortium [23], which is an initiative of major European car manufacturers and suppliers. Furthermore, a test bed for functional tests and demonstrations is implemented which will be developed further on toward a reference system for the Car2Car Communication Consortium specifications.

1.2.5.1 System Implementation

The NOW project has implemented a software prototype of the developed system covering radio, networking and applications. The radio subsystem implements IEEE 802.11 physical and MAC layer based on commercial WLAN chip-sets and the MADWIFI multi-mode software driver. For IEEE 802.11 p compatibility, the driver is significantly enhanced, including extensions to operate at the protected 5.9 GHz frequency band, control of selected radio parameters on a per-packet basis from the network layer and exchange of signaling data between the MAC and upper protocol layers. The communication system is mainly developed in C for the Linux operating system. Applications are implemented in Java/OSGI.

1.2.5.2 Research and Experiments

- *Safety information dissemination:* The NOW project has developed a hybrid scheme of network-layer and application-layer forwarding [24–26]. The network layer protocol provides a sender-oriented and Geo-addressed distribution of data packets (a mechanism capable to efficiently distribute a message to all nodes inside an area) based on traditional packet-switching concepts. Applications enable a receiver-oriented scheme for dissemination, in which every node decides individually about information re-broadcasting. The latter approach enables flexibility as well as aggregation and modification of the information carried in the message payload. The combination of both schemes results in a hybrid approach, which enables rapid distribution of data packets by Geocast and adaptive dissemination of information.

1.3 Empirical VANET Studies at SJTU

In this book, we will elaborate major VANET studies conducted at Shanghai Jiao Tong University (SJTU) based on large-scale realistic vehicular traces. The reminder of this book is organized as follows:

In Chap. 2, we first introduce the ShanghaiGrid (SG) project from which we have collected GPS traces of more than 6,850 taxis and 3,620 buses. Based on those traces, we study VANET topics ranging from realistic mobility model and opportunistic data forwarding protocols to real-time vehicle tracking and traffic and environment sensing applications. We then present the details of collecting those data and main challenges in data processing.

In Chap. 3, we present the realistic mobility model study which aims to reveal the fundamental characteristics of vehicular mobility in urban environments and to establish simple yet effective mobility models for new routing algorithm design and realistic simulations. Based on the comprehensive analysis on the distribution of time intervals between two consecutive contacts between a pair of vehicles (called inter-contact time or ICT), we find that vehicles can "meet" very frequently with each other, which can greatly facilitate data communications. Specifically, the complementary cumulative distribution function (CCDF) of ICT between the same pair of vehicles exhibits an exponential decay. Furthermore, we also point out that the major reason for this phenomenon is caused by the layout of road networks and popular places (called "traffic influxes") existing on the itineraries of vehicles.

In Chap. 4, we describe two opportunistic data forwarding protocols, which provide mechanisms for a vehicle to deliver a packet over the vehicular ad hoc network utilizing the communication opportunities among neighboring vehicles. Based on the real vehicular trace data collected in SG, the time and duration of contacts as well as the social connections between any pair of experimental vehicles are analyzed. We find that the ICT between a pair of vehicles has apparent temporal correlation, which is utilized to design a new opportunistic data forwarding algorithm. The core idea of this algorithm is to choose a better candidate with a shorter expected contact time with the destination as the next data relay. Furthermore, we also find that vehicles can form apparent social structures by aggregating individual pairwise contacts. Inspired by this observation, we propose an innovative data forwarding algorithm, which leverage both contact-level and social-level vehicular mobility to improve the performance. With those algorithms, the end-to-end delay and network traffic cost can be largely reduced whereas the delivery ratio can be improved as well.

In Chap. 5, we present the real-time vehicle tracking service in the SG project, which refers to tracking the current position of a vehicle in real time. In the system, a vehicle attached with an active RFID tag or a WiFi wireless card can be captured by local nodes (associated with several RFID readers or WiFi APs) largely deployed as infrastructure. By enquiring these largely-distributed nodes, any system-enabled vehicle can be localized and tracked in real time. The biggest challenge in implementing this service is to guarantee the quality of service in terms of response time and meanwhile to minimize the network cost caused by location information updating and query processing. To tackle this difficulty, a novel distributed scheme, called HERO, is devised. In HERO, local nodes are typically deployed at intersections and interconnected according to the geographical positions as a backbone network. As vehicles pass by, its location information can be locally captured and stored. By organizing those local nodes into a well-designed hierarchical structure, a query injected in the network from any local node can be forwarded to the node which has the latest location information of the vehicle within a given query latency requirement. In addition, location information updating aroused by the movements of the target vehicle is also restricted only within a limited area. In this way, HERO can achieve real-time query response time and minimize overall network traffic as well.

In Chap. 6, we describe the urban traffic condition perception service in SG, which refers to determine the traffic condition on urban surface roads based on instant GPS speed reports collected from experimental vehicles. Due to concrete jungles in the urban settings, the location information obtained from GPS reports often contains errors. In addition, those traffic sensory data are very sparse in terms of temporal and spatial distribution. How to accurately estimate traffic condition based on the coarse data set is very challenging. To realize this service to the public, an MSSA-based scheme is implemented, where the estimated traffic condition on a certain road segment is further treated as a time series with missing points. MSSA is used to fill up those missing points and remove "noise" part contained in the data.

Chapter 2
Dealing with Vehicular Traces

2.1 Introduction to the Shanghai Grid Project

Intelligent transportation systems (ITSs) [72–74] have been evolving rapidly in the past two decades, leveraging advanced computing and communication technologies. ITSs help coordinate traffic condition, improve safety, reduce environmental impact, and make efficient use of available resources. Shanghai, the largest metropolis in China, covers an area of 5,800 km^2 and has a large population of 18.7 million. The economy of Shanghai is soaring today and the growing traffic has become a serious challenge. In response to the challenge and the needs of the public, the Shanghai government has established the SG project cooperated with SJTU since 2005, with the ambitious goal of building a metropolitan-scale traffic information system. The goals of the project are twofold. First, it tries to make the available transportation infrastructure to be used more efficiently. Second, it aims to provide the public with a wide spectrum of ITS applications, ranging from real-time traffic information, trip planning and optimal route selection, to congestion avoidance and bus arrival prediction.

In this chapter, we first introduce three vehicular trace data sets involving tens of thousands of public vehicles collected from the SG project and from Shenzhen, another metropolis in south China. The reason that we collect these data is mainly to better understand vehicular mobility and to conduct informed design of message forwarding algorithms between vehicles. Then, we present the main challenges encountered to process those data for future VANET studies.

2.2 Collecting Vehicular GPS Traces

In the SG project, each experimental vehicle is deployed with a GPS unit and a GPRS wireless communication module. As such a vehicle runs along the roads in the city, it periodically sends a GPS report back to a data center via a GPRS channel. Due to the GPRS communication cost for data transmission, reports are

H. Zhu and M. Li, *Studies on Urban Vehicular Ad-hoc Networks*,
SpringerBriefs in Computer Science, DOI: 10.1007/978-1-4614-8048-8_2,
© The Author(s) 2013

Fig. 2.1 A taxi with a commercial GPS device installed (*highlighted* in the inset), the location and operational information thus can be periodically sent back via GPRS wireless channels

usually sent at rather large intervals, typically once per minute. We have collected three datasets consisting of GPS traces of buses and taxies from two cities in China:

Shanghai Taxies: We collected the GPS trace of taxies in Shanghai collected between Feb 1 and Mar 3, 2007. We chose 2,109 taxies in the datasets which have consecutive GPS reports on each day during the 31 days. The specific information contained in such a report includes: ID, the longitude and latitude coordinates of the current location, timestamp, moving speed, and heading direction. In addition, the information contained in a taxi GPS report also reports whether passengers are onboard. The granularity of reports is 1 min for taxies with passengers and about 15 s for vacant ones. Figure 2.1 illustrates an experimental taxi in Shanghai. In Fig. 2.2, the geometry of destinations of all taxi deliveries on Shanghai map during Feb of 2007 is shown, where every colored dot presents the average number of destinations per taxi per day located in the corresponding 300 m × 300 m square area on the map.

Shanghai Buses: The trace consists of GPS reports sent from 2,501 buses which serve on 199 routes and cover the main downtown area (within Neihuan Viaducts about 120 km^2) between Feb 19 and Mar 5, 2007. Figure 2.3 shows the coverage of all experimental bus routes. A commuting bus periodically sends GPS reports back to a backend data center via GPRS channel. The information contained in a report is similar to that of taxies except that there are more fields contained, such as whether the bus is arriving at a stop or a terminal is also sent. Due to the GPRS communication cost for data transmission, reports are sent at a granularity of around 1 min.

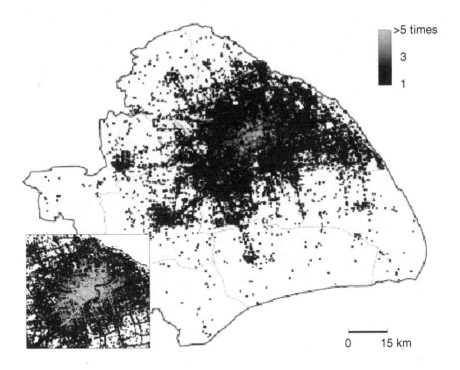

Fig. 2.2 The geometry of destinations of all taxi deliveries on Shanghai map during Feb 2007. Every *colored dot* presents the average number of destinations per taxi per day located in the corresponding 300 m × 300 m square area on the map

Fig. 2.3 The distribution of bus lines within the downtown area of Shanghai city, with 199 bus lines denoted by *red lines*

Shenzhen Taxies: We also collected the GPS trace of taxies in Shenzhen in October, 2009. The data format is similar to that of Shanghai taxi trace. We chose 8,291 taxies which continuously send GPS reports during the whole period. Taxies in Shenzhen always send GPS reports on every 1 min. Figure 2.4 demonstrates the

Fig. 2.4 The geographical
distribution of GPS reports
from all experimental taxies
in Shenzhen on Oct 1, 2009

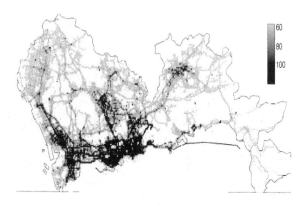

Table 2.1 Comparison of three data sets

Data set	Shanghai bus	Shanghai taxi	Shenzhen taxi
Number of vehicles	2,501	2,109	8,291
From date	Feb 19, 2007	Feb 1, 2007	Oct 1, 2009
Duration (day)	15	31	31
Granularity (second)	60	15^*, 60^{**}	60
Number of contacts	1,229,380	22,053,178	23,968,860
Mean ICT (minute)	31.8	47.6	30.5

* Vacant
** Passengers onboard

geographical distribution of GPS reports from all experimental taxies on October 1, 2009, where every colored dot presents the average number of reports per taxi per day located in the corresponding 300 m × 300 m square area on the map.

We choose taxies and buses to study for two reasons. First, taxies and buses shows two distinct mobility patterns, namely, rather random and well scheduled, respectively. Second, the privacy problem is less concerned since we use public vehicles. As privacy preservation schemes progress and more mobility data of private vehicles available, it is invaluable to study private vehicles in the future. Key statistics of the traces are listed in Table 2.1.

2.3 Challenges and Issues in Data Analysis

To study VANETs in urban scenarios, it is ideal to collect GPS reports for a sufficient long period of time from various types of vehicles 24 h a day with a granularity measured in seconds. In practice, due to the deployment and communication costs and privacy issues, we only collect GPS reports from public vehicles, i.e., taxies and buses with a granularity measured in about 1 min. As a result, the data sets are very sparse in terms of temporal and spatial distributions.

Fig. 2.5 CDFs of GPS sample density at each road

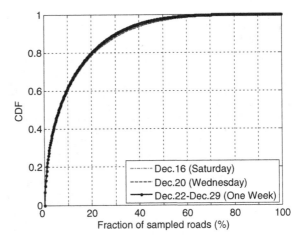

We first examine the geographic distribution of GPS data. For example, from Fig. 2.2, it can be seen that most of the GPS samples are scattered in the downtown area where taxies congregate more densely than in suburbs. The cumulative distribution functions (CDF) of sample density on each road are shown in Fig. 2.5. The data are taken on a weekend, on a workday and for a whole week, respectively. We observe an obvious Pareto distribution in which the "80-20 rule" [27] stands (i.e., 20 % of the road segments owns 80 % of the data).

We then examine the distribution of taxi GPS data in time dimension. We are interested in the probability distribution of the inter-report times, which refers to the time intervals between any two consecutive reports received from a location over time. Figure 2.6 shows the complementary cumulative distribution function (CCDF) of inter-report times. It can be seen that the middle part of the CCDF is almost linear in log–log scale, which indicates a power law. This means a location may frequently has no sensory data for a long time. Figure 2.7 shows the CCDFs

Fig. 2.6 CCDF of inter-report times

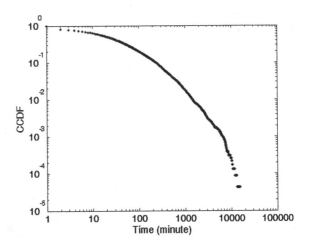

Fig. 2.7 CCDFs of the proportion of time with no sensory data

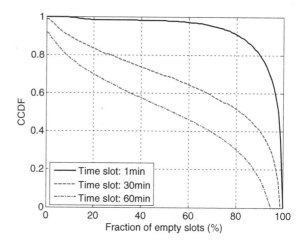

of the proportion of time with no sensory data in a day in different observation granularities. The time windows used to collect sensory reports are 1, 30 and 60 min, respectively. It shows that about 90 % of roads have no samples in 80 % of the 1,440 min in a day. The fraction is about 50 % when counting the number of road segments that are short of samples for 12 h in a day.

Besides the sparseness, the collected GPS trace data are also erroneous with noise. In the city setting with dense high buildings and viaducts, the error of GPS reports from taxies can be as large as 100 m. To tell which road a taxi is actually monitoring, we need to recover each sample back on track. We deal with this problem using our map-matching algorithm. More specifically, the algorithm prefers those roads with minimum projection distance from the report and minimum angle deviation between the heading direction of the taxi and the road. This simple yet effective strategy works well in most situations and can gain very high accuracy compared with real itineraries. In more complicated cases where the

Fig. 2.8 CDF of speed difference at the same location at the same time

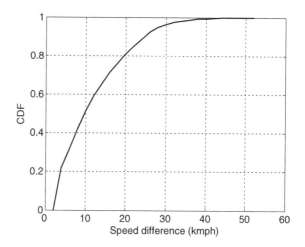

geographical distance between these two consecutive records could exceed thousands of meters, we need to consider the mobile context of the taxi. The algorithm examines several previous and successive reports to determine the most possible road segment where the report issued. Our on-road experiment results show that our map-matching algorithm can reach about 98 % accuracy with the left regarded as an inevitable source of noise.

In addition, we also find that individual reports vary significantly even they are collected from the same location at the same time. Figure 2.8 shows the CDF of speed difference derived from reports at the same location at the same time. It can be seen that the CDF increases slowly with a relatively long tail, which implies the individual reports can vary largely. The derivation of this variance may be ascribed to individual driving behavior. For example, a taxi may stop arbitrarily to pick up or drop passengers. In other words, each sensory data report is associated with a certain degree of noise. Despite these inaccuracies, the GPS trace data are very valuable to study VANETs since they cover thousands of vehicles and last for 1 month.

Chapter 3
Realistic Vehicular Mobility Models

3.1 Introduction to Mobility Models

In VANETs, vehicles equipped with wireless communication devices can transfer data with each other (vehicle-to-vehicle communications) as well as with the roadside infrastructure (vehicle-to-roadside communications). In order to successfully transfer data from a vehicle to another, the vehicle needs to first wait until it geographically meets other vehicles (within the communication range of each other) for data-relay. Applications based on this type of data transfer will strongly depend on vehicular mobility characteristics, especially on how often such communication opportunities take place and on how long they last. In this chapter, we focus on studying the metric called inter-contact time [28–30], which denotes the time elapsed between two successive contacts of the same two vehicles. Since data transfer arises in a store-carry-forward fashion, the inter-contact time (ICT) of the two vehicles is a major component of the end-to-end delay, as it presents how long it takes to encounter the other mobile vehicle to have any chances to forward/relay the data for communications. Larger inter-contact time results in larger end-to-end delay.

In the literature, bunches of studies have made their effort on revealing the relationship between the underlying mobility models of nodes and the consequent characteristics of the inter-contact time in MANETs. In general, these studies can be classified into two categories: theoretical mobility models based and empirical trace based.

3.1.1 Theoretical Mobility Models

Most of existing studies focus on theoretical models such as random walk mobility models (RWM) [31–33], random waypoint mobility models (RWP) [34–36] and random direction mobility models (RDM) [37].

H. Zhu and M. Li, *Studies on Urban Vehicular Ad-hoc Networks*,
SpringerBriefs in Computer Science, DOI: 10.1007/978-1-4614-8048-8_3,
© The Author(s) 2013

For example, in RWM, a mobile node moves from its current location to a new location by randomly choosing a direction and speed in which to travel. The new speed and direction are both chosen from pre-defined ranges, respectively. Each movement in the RWM occurs in either a constant time interval or a constant distance traveled, at the end of which a new direction and speed are calculated. If a mobile node which moves according to this model reaches a simulation boundary, it "bounces" off the simulation border with an angle determined by the incoming direction. The mobile node then continues along this new path. Many derivatives of RWM have been developed including the 1-D, 2-D, 3-D, and n-D walks. The Random Walk Mobility Model is a widely used mobility model, which is sometimes also referred to as Brownian motion. It is clear that RWM is a memoryless mobility pattern because it retains no knowledge concerning its past locations and speed values, which means the current speed and direction of a mobile node is independent of its past speed and direction. This characteristic can generate unrealistic movements such as sudden stops and sharp turns. Similarly, in RWP, a mobile node starts by staying in one location for a certain period of time (i.e., a pause time). Once this time expires, the mobile node chooses a random destination in the simulation area and a speed that is uniformly distributed between pre-defined ranges. The mobile node then travels toward the newly chosen destination at the selected speed. Upon arrival, the mobile node pauses for a specified time period before starting the process again. Using RWP should notice the initial positions of mobile nodes in a simulation. Randomly chosen positions could cause high variability of a mobile node in connectivity with neighboring mobile nodes [41].

A majority of research results have uncovered a common property of many theoretical mobility models that the tail of the ICT distribution decays exponentially. In other words, for these models, the ICT is light tailed. For example, authors [33, 38, 39] draw this conclusion through numerical simulations based on RWP mobility models. Furthermore, some theoretical results show that the first and second moments of the inter-contact time are bounded above under Brownian motion model on a sphere. In particular, authors in [40] rigorously prove that a finite domain is one of the key aspects in creating the exponential ICT tail distribution. This is because finite boundaries actually force mobile node to move only within a certain region and hence increase the meeting opportunities between nodes. While theoretical mobility models facilitate problem analysis, they are far beyond reality and not practical in designing networking protocols for real systems and their performance analysis.

3.1.2 Empirical Mobility Models

In recent years, there has emerged more research work taking experimental study on the characteristics of the inter-contact time. For example, some empirical results [28–30] based on human mobility show that the tail distribution of the inter-contact time is far from being exponential, but can be approximated or lower bounded by a power law. These results are based on real traces such as human contacts while at conferences [30], campus WiFi login records [29, 42] and a Bluetooth network containing hundreds of people in an office [30].

It is apparent that the mobility of vehicles is significantly different from that of human beings in terms of speed, constraints of road transportation systems and travel distance. Although these empirical results based on human mobility depict another scene of the ICT distribution, the situation in vehicular environments is still left unknown. There have been some pilot projects setting up to study vehicular mobility. For example, DieselNet [15] at UMass consisting of 40 buses studies the aggregated inter-contact time distribution at a granularity of bus route and find a clear periodic structure in the inter-contact times between two bus routes. For the lack of enough contact samples between two individual buses, the bus trace data are not sufficient for studying the distribution of the inter-contact time between two individual buses. In the RAPID routing protocol [43], it is assumed that the distribution of bus inter-contact times is exponential to make their problem tractable.

3.2 Empirical Vehicular Mobility Analysis

In order to have a better understanding of practical constraints in opportunistic data transfer between vehicles, experiments involving thousands of vehicles over a long time span of months are in pressing demand. In practice, the deployment cost for such large-scale experiments would be enormous. In this chapter, we study the vehicular mobility by analyzing the large volume of trace data that we have described in Chap. 2. We are interested in how often transfer opportunities can occur between vehicle pairs since it is the key factor that impacts the end-to-end delay for data delivery in VANETs. We first describe the method to extract contacts between each pair of experimental vehicles from the trace. Then we present the inter-contact time characteristics embedded in the trace data. Finally, we discuss the possible reasons behind our key observation on the vehicular ICT distribution.

3.2.1 Extracting Inter-Contact Times

Ideally, all connection opportunities encountered 24 h a day, with a granularity measured in seconds should be recorded in the data for study. Since we collect GPS reports in discrete time, we use a sliding time window to check contacts between a pair of taxies. Here we make the assumption that two vehicles would be able to communicate (called a *contact*) if their locations reported within the time window are within the communication range. For the example if Fig. 3.1, suppose we have two real contacts C_1 and C_2 happening between vehicle v_1 and v_2. As v_1

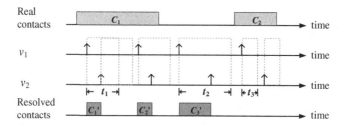

Fig. 3.1 Extract contacts from GPS reports of vehicle v_1 and v_2. Boxes in *dotted line* denote sliding time windows of different granularities used to check contacts. Individual GPS reports are presented by *short arrow line* segments

and v_2 keep sending reports, given a communication range, we can slide a time window along the time axis to check contacts.

The above assumption, though, can introduce inaccuracies in the following two cases.

First, if a relatively large time window is used, we may introduce false contacts into consideration. This is because two taxies may have already run far away from their reported locations. Therefore, the retrieved contact may never really exist. For example, in Fig. 3.1, we may get a false contact C_3' if a large time window t_2 is used even though there is no real contact at all. The consequence of introducing false contacts is that it increases the weight of small values of inter-contact times in the distribution since these false contacts cut large ICTs into small pieces.

Second, if a small time window is used, we may omit real connection opportunities. This is because two taxies might indeed have a contact but did not send out reports simultaneously. In this case, we may not capture this contact due to the small size of the time window. This is the case of contact C_2 as shown in Fig. 3.1. The consequence of omitting real contacts is that it causes large values of ICTs since two real small ICTs are now considered as a single huge one. Moreover, using small time windows to check contacts can add the weight of small values into the ICTs distribution. For example, in Fig. 3.1, we will get C_1' and C_2' rather than the real one, C_1, when we take t_1 to check contacts. To eliminate this effect of using a small time window, we calculate the correlation between two contacts with a small ICT. Specifically, given the reported locations and speeds of each taxi, we calculate the remaining contact time of the first contact as the time these two taxies move along the same directions and at the same speeds before they are out of the communication range. If the second contact is contained within the remaining contact time of the first contact, we make a decision that these two contacts should be merged into one. Vice versa, we can infer when the second contact started and further check whether the first contact can be merged.

Despite these inaccuracies, the GPS trace data are very valuable to study vehicular mobility models since they cover thousands of vehicles and last for one month. In addition, as most of the GPS reports are sent at a relatively small period (48 s on average), the deviation of the computed distance of two reported locations within such a small time window from the actual distance between two taxies is small.

We refer to ICT as the time elapsed between two successive contacts of the same vehicles as defined in [28–30]. Specifically, the ICT is computed at the end of each contact, as the time period between the end of this contact and the start of the next contact between the same two vehicles. It should be noted we do not take into consideration the ICT starting after the last contacts.

3.2.2 Identifying Exponential ICT Tail

We plot the ICT distribution for the selected trace data in Fig. 3.2. The distribution of ICT is computed among all pairs of 2,109 Shanghai taxies during the whole February in 2007. We get six different sets of ICTs by combining different communication ranges and time window sizes used in the contact extraction. The time windows are set to one second, thirty seconds and one minute, respectively, accompanied with two communication ranges of 50 and 100 m. All plots describe the tail distribution function, i.e., $P\{X > t\}$, in linear-log scale.

The most interesting part in Fig. 3.2 is that all plots exhibit a very clear *exponential tail*, i.e., $P\{X > t\} \sim e^{-\beta t}$ [44]. This can be indicated by the fact that all plots are almost straight lines with different negative slopes in linear-log scale, from the very beginning of time and over a large range of timescale. This implies that, to some extent, taxies can frequently meet with each other. Thus, data delivery via vehicle-to-vehicle communications would experience smaller end-to-end delay. Besides the exponential parts, we also notice that, gradually, all six distributions start to deviate from the exponential decay and drop faster till the end. This rapid cutoff is caused by the limited duration of the trace data, i.e., one month in our experiment. The reason is that ICTs that last longer than the duration of the

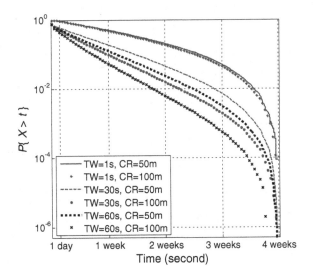

Fig. 3.2 Tail distribution of ICT: data collected from 2,109 taxies in Shanghai city during the whole February of 2007. Contacts are collected under three different time windows of one second, thirty seconds and one minute and two communication ranges of 50 and 100 m

trace data cannot be observed and those ones with very large values close to the duration are less likely to be found. The effect of observation duration has also been noted in Augustin Chaintreau et al.'s study based on human mobility [28].

To identify the exponent constant β, we perform the least-square regression analysis to the extracted ICT. It should be noted that the cutoff part of data should not be used for regression as they can be considered as artifacts to the tail distribution. More specifically, we separate the whole regression processing into two steps. First, we need to identify the divide point from which the tail distribution function stops exponential decay. This can be achieved by seeking for the point from which the second derivatives (decay acceleration) of the log-scaled $P\{X > t\}$ are nonzero. Then, we apply polynomial regression to the log-scaled $P\{X > t\}$ over the range from the first point to the divide point. Figure 3.3 shows the regression result on the lowest plot in Fig. 3.2.

3.2.3 Recognizing Traffic Influxes

The surprising finding of the exponential decay on the ICT tail distribution is in a sharp contrast to several recent empirical results on the ICT based on extensive human mobility traces [28–30]. These results indicate that the tail behavior of the ICT can be approximated or lower bounded by a power law, i.e., $P\{X > t\} \sim t^{-\alpha}$, for some constant $\alpha > 0$. More spectacularly, it was shown that the power law exponent α is normally less than one, making the expected end-to-end delay tend to be infinite, independent of any forwarding algorithm, if the network only contains a finite number of devices. We claim that the tail distribution of the ICT based on vehicular mobility satisfies an exponential decay or a light tail. An exponential decay means the tail distribution function decreases rapidly over this range. For example, in the lowest plot in Fig. 3.2, about 45 % of ICTs are greater than one day, and only 5 % are greater than one week.

Fig. 3.3 Tail distribution of the ICT under the time window of one minute and the communication range of 100 m is very well approximated by an exponential-like distribution $y = e^{-\beta \cdot t}$ with $\beta = 3.71 \times 10^{-6}$. We omit artifact part in the observed data by examining the nonzero decay acceleration in the tail distribution plot

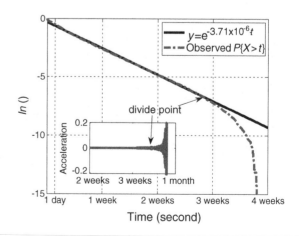

The discrepancy between human mobility and vehicular mobility described above calls for answers to the following question: *What is the key factor that makes vehicles meet with each other frequently?* The answer to this question is of great importance since finding the essential reasons will provide fundamental guidelines to many related studies in VANETs, such as capacity-delay trade-offs and design and performance analysis of data forwarding algorithms based on exponential ICT. Recently, authors in [40] prove that, in any RWP model and any RWM model, the tail of the ICT between two independent mobile objects decays at least exponentially fast as long as the boundary is finite. This result can be easily understood because, in a finite region, while hitting on the boundary, a mobile node changes its current status correspondingly according to the type of the boundary. For example, if the boundary is "reflective", the node will be bounced back into the region with a reflective moving direction and speed and hence increases contact opportunities with other objects. The boundary effect, however, is not the major reason in our case where vehicles rarely run out of the city.

By examining the geographical distribution of destinations of loaded taxies, we find out that the destination distribution over the city is quite uneven. This is shown in Fig. 2.2 where it can be seen there are many hot areas scattered in the downtown area, which attracts a large amount of traffic. It can be seen there are many hot areas scattered in the downtown area. The inset shows aggregated GPS report distribution (destinations not involved). We call such an area a *traffic influx* if traffic tends to converge around this area. The reason of traffic-influx-forming can be very complicated in urban settings. For example, those areas of interest such as large business/commercial centers or high-density residential districts are very likely to be traffic influxes since they keep attracting people to gather. Furthermore, in the road transportation systems, there are many major traffic junctions connecting a high volume of traffic which may also have the same effect (shown in the inset of Fig. 2.2). In addition, bad or inappropriate design of road networks may also form another kind of traffic influxes where constant traffic congestion occurs. Intuitively, a traffic influx has the effect of gathering vehicles from time to time and hence enormously increases contact opportunities of vehicles.

With these findings, we make further remarks as follows. First, this implies that most of the taxies perform deliveries only within the city. We argue that a taxi spontaneously presents itself in popular areas most of the time rather than being passively "reflected" back into these areas when hitting certain physical district/city boundaries [40]. Therefore, boundary effect is not the major reason of generating the exponential tail. Second, even taxies which perform arbitrary deliveries do have special mobility patterns as opposite to random mobility. A loaded taxi has an explicit destination instead of randomly-picked one. Third, it is clear to see that these destinations are densely congregated forming hot areas in the city. The effect of hot areas is to gather vehicles all the time and hence can enormously increase their contact opportunities.

3.3 Modeling Urban Vehicular Mobility

In this section, we take a penetrating study on the impact of traffic influxes from an analysis perspective. We first give some basic definitions and preliminaries related to our analysis. Then we introduce previous theoretical results based on RWM mobility models for the purpose of comparison. Last, we present our mobility model and major theoretical results.

3.3.1 Definitions and Preliminaries

We consider two arbitrary vehicles v_1 and v_2, each of which moves according to some mobility model in an infinitely large region Ω (i.e., $\Omega = R^2$). Let $V_{v1}(t)$, $V_{v2}(t) \in \Omega$ be the position of the vehicles v_1 and v_2 at time t, respectively. We assume that $V_{v1}(t)$ and $V_{v2}(t)$ are independent and two vehicles can communicate with each other whenever they are within the communication range R_c. Due to the interference and signal loss of wireless links, two vehicles within communication range may still not be able to perform real data transfer. Since we focus on the characteristics of the tail behavior of the ICT, we study the potential communication opportunities between two vehicles and leave the successful data transfer rate with no further discussion.

Definition 3.1 The ICT T_I of vehicles v_1 and v_2 is defined as

$$T_I \triangleq \inf_{t > 0} \{t : \|V_{v1}(t) - V_{v2}(t)\| \leq R_c\} \tag{3.1}$$

given that $\|V_{v1}(0) - V_{v2}(0)\| = R_c$ and $\|V_{v1}(0^+) - V_{v2}(0^+)\| > R_c$. Here, $\|\cdot\|$ is the two-dimensional Euclidian distance.

Let X be a random variable that has Gaussian distribution with mean μ and variance σ^2, i.e., $X \sim N(\mu, \sigma^2)$.

If $X \sim N(\mu_X, \sigma_X^2)$ and $Y \sim N(\mu_Y, \sigma_Y^2)$ are two independent random Gaussian distributed variables, we have

Property 3.1 *The difference of X and Y is also Gaussian distributed,*

$$U = X - Y \sim N(\mu_X - \mu_Y, \sigma_X^2 + \sigma_Y^2) \tag{3.2}$$

3.3.2 RWM Without Traffic Influx

For the sake of comparison, we examine the tail behavior of the ICT under a two-dimensional discrete-time isotropic RWM mobility model [3, 12] with no traffic influxes existing in the experimental region. In this model, at the beginning of each time slot t, a vehicle randomly selects a direction uniformly from $[0, 2\pi]$ and moves a random distance D which is chosen from $(0, \infty)$ at a random speed chosen from pre-defined $[v_{\min}, v_{\max}]$. After it reaches the destination, it repeats the same procedure and starts the next run. We introduce the following result:

Theorem 3.1   (Theorem 4 in [12]): *Suppose that two independent vehicles v_1 and v_2 move according to the two-dimensional isotropic random walk mobility model described above. Then, the ICT T_I of vehicles A and B is $P\{T_I > t\} \geq C.t^{-1/2}$, where C is a constant, when t is sufficiently large.*

3.3.3 Mobility Model with Traffic Influx

With the constant observation of vehicle-gathering in urban environments, we introduce a mobility model to characterize the gathering effect at a traffic influx.

In our model [45], a vehicle randomly walks in an infinitely large region. During the process, it revisits a fixed location, called the *traffic influx*, at least once within a given time period T. More specifically, at the beginning of time, a vehicle randomly selects a direction uniformly from $[0, 2\pi]$, a random distance D chosen from $(0, \infty)$ and a speed randomly chosen from pre-defined $[v_{\min}, v_{\max}]$. Besides, it also sets up a TIMER with the value of T. Then the vehicle starts to conduct random walk until the TIMER has expired. In that case, the vehicle stops randomly walking and heads for the traffic influx. Once it reaches the traffic influx, the vehicle resets its time and repeats the whole process. We call T the vehicle's maximum periodicity with respect to the traffic influx. The magnitude of T thus indicates how attractive this traffic influx is to this vehicle. A small T means the vehicle very often appears at the traffic influx. As T increases to infinity, chances are that the appearance of the vehicle at the traffic influx is prolonged. In this model, each vehicle can take its own maximum periodicity. Figure 3.4 illustrates the mobility model with traffic influx.

3.3.4 Analysis on ICT Under Mobility Model with Traffic Influx

We present following assumption and lemma.

Assumption 3.1   At any time t, the mobility of each vehicle is independent, i.e., a vehicle chooses its destinations and routes according to the mobility model and its own preferences to the traffic influx.

Lemma 3.1   *Given a specific maximum periodicity T of a vehicle, the region where the vehicle moves can be upper bounded by a disk with a radius $r = v_{\max}.T$, where $v_{\max}$ is the maximum velocity of the vehicle.*

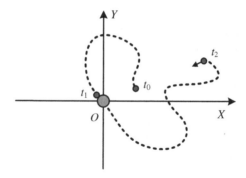

Fig. 3.4 A vehicle v_1 moves according to the mobility model with traffic influx. The *red dot* located at the origin denotes the traffic influx. At time t_0, v_1 sets up a timer and starts to randomly walk. At time t_2, due to the expiration of its timer, the vehicle terminates current random walk and return to the traffic influx. Note that before the expiration of the timer, it is possible for v_1 to be present at the traffic influx with random walk (e.g., at time t_1)

Proof Consider the extreme case where the vehicle starts from the traffic influx and keeps moving without changing its direction. Since the vehicle is bound to return to the traffic influx before its maximum periodicity T, the maximum distance the vehicle can run away from the traffic influx is $v_{max}.T$. In addition, because the vehicle can choose any direction uniformly from $[0, 2\pi]$, the reachable area will be covered by a disk with the radius $r = v_{max}.T$.   ∎

The effect of T on a vehicle is actually equal to putting a constraint on the maximum area the vehicle can reach. Obviously, a larger T allows the vehicle to move further in the region.

Lemma 3.2 *Under the mobility model described above, the contact opportunities of a vehicle can be lower bounded by any mobility model that yields independent symmetric Gaussian location distribution of vehicles.*

Proof Since a vehicle randomly selects a direction while performing random walk, the probability density function (PDF) of the location distribution of the vehicle is thus independent in independent in orthogonal directions and symmetric as well. Moreover, with the existence of T, the vehicle presents itself more frequently around the traffic influx, which makes the PDF yield a convex shape with the highest probability appearing at the traffic influx. The PDF in x-axis is shown as P(x) in Fig. 3.5.

Fig. 3.5 The contact opportunities on the PDF $P(x)$ can be lower bounded by the contact opportunities on a Gaussian distribution $N(0, \sigma^2)$ with $\sigma = v_{max}.T$

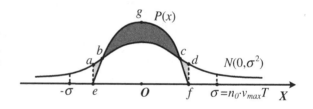

We now prove that a mobility model yielding independent symmetric Gaussian location distribution of vehicles will satisfy Lemma 3.2, as long as the peak of the bell shape of the Gaussian is located at the traffic influx and the probability at the peak is smaller than $P(x = 0)$.

Since two vehicles constantly revisit the common traffic influx located at the origin in Fig. 3.5, each of them presents their selves more frequently when they are close to the traffic influx and therefore have more contact opportunities than when they are far away from it. Therefore, the contact opportunities can be determined by the area of $P(x)$. Because the probability at the peak of the Gaussian is smaller than $P(x = 0)$, we have $S_{abe} + S_{cdf} < S_{boc}$, where S_{abe}, S_{cdf} and S_{boc} denote the area of region abe, cdf and boc, respectively, as shown in Fig. 3.5. This is easy to be verified since $S_{boc} + S_{bcef} = 1$ (i.e., the cumulative density function of $P(x)$ is 1) whereas $S_{abe} + S_{cdf} + S_{bcef} < 1$ (any truncated area of Gaussian PDF is less than 1). This completes the proof. ∎

We now present our major result as follows:

Theorem 3.2 *For any two arbitrary vehicles moving according to the mobility model described above, there exists a constant c such that $P\{T_I > t\} \leq e^{-ct}$ for all sufficiently large t as long as they have finite maximum periodicities.*

Proof Suppose the maximum periodicity of vehicles v_1 and v_2 are T_1 and T_2, respectively. According to Lemma 3.2, we arbitrarily choose a mobility model for v_1 that satisfies the condition of Lemma 3.2. The variances of the Gaussian location can be expressed by $\sigma_{X1}^2 = \sigma_{Y1}^2 = (n_1 v_{max} T_1)^2$, where n_1 is a positive number. Similarly, the variances of the Gaussian location for v_2 can be expressed by $\sigma_{X2}^2 = \sigma_{Y2}^2 = (n_2 v_{max} T_2)^2$.

Let $P_\Lambda\{T_I > t\}$ be the probability that v_1 never meets v_2 until time t under corresponding Gaussian distribution and $P_\Xi\{T_I > t\}$ be the probability that v_1 never meets v_2 until time t under our mobility model. The condition described above will guarantee $P_\Xi\{T_I > t\} \leq P_\Lambda\{T_I > t\}$. Next we will prove that there exists a constant c such that $P_\Lambda\{T_I > t\} = e^{-ct}$.

Due to the independency and the same variances in each dimension, we can express the PDF of Gaussian distribution in each dimension as follows:

$$f_X(x) = \frac{1}{\sqrt{2\pi}\sigma} \exp\left(-\frac{x^2}{2\sigma^2}\right) \tag{3.3}$$

$$f_Y(y) = \frac{1}{\sqrt{2\pi}\sigma} \exp\left(-\frac{y^2}{2\sigma^2}\right) \tag{3.4}$$

Thus, the joint PDF $f(x, y)$ is

$$f(x,y) = f(x)f(y) = \frac{1}{2\pi\sigma} \exp\left(-\frac{x^2+y^2}{2\sigma^2}\right). \tag{3.5}$$

Note that the relative position of v_2 with reference to v_1 in each dimension is also independent. With Property 3.1, we have

$$f_{X_1-X_2}(x) = \frac{1}{\sqrt{2\pi} \cdot \sqrt{\sigma_1^2 + \sigma_2^2}} \exp\left(-\frac{x^2}{2 \cdot (\sigma_1^2 + \sigma_2^2)}\right) \tag{3.6}$$

and

$$f_{Y_1-Y_2}(y) = \frac{1}{\sqrt{2\pi} \cdot \sqrt{\sigma_1^2 + \sigma_2^2}} \exp\left(-\frac{y^2}{2 \cdot (\sigma_1^2 + \sigma_2^2)}\right). \tag{3.7}$$

For clarity of writing, we drop the index symbol $x_1 - x_2$ and $y_1 - y_2$ simply write the joint PDF as

$$f(x,y) = \frac{1}{2\pi \cdot (\sigma_1^2 + \sigma_2^2)} \exp\left(-\frac{(x^2 + y^2)}{2 \cdot (\sigma_1^2 + \sigma_2^2)}\right) \tag{3.8}$$

At any time t, the probability that v_1 and v_2 does not meet is the probability that they are out of the communication range R_c. Let E denote the event that two vehicles do not meet at any time t. Denote by $r = \sqrt{x^2 + y^2}$ the distance between the two vehicles at time t. We have

$$\begin{aligned}
P_\Xi\{E\} &= P_\Xi\{\|V_1(t) - V_2(t)\| > R_c\} \\
&= \frac{1}{2\pi(\sigma_1^2 + \sigma_2^2)} \iint_{x^2+y^2 > R_c^2} \exp\left(-\frac{1}{2}\left(\frac{x^2 + y^2}{\sigma_1^2 + \sigma_2^2}\right)\right) dxdy \\
&= \frac{1}{2\pi} \int_0^{2\pi} d\theta \int_{R_c}^{+\infty} \frac{r}{(\sigma_1^2 + \sigma_2^2)} \exp\left(-\frac{r^2}{2 \cdot (\sigma_1^2 + \sigma_2^2)}\right) dr \\
&= -\exp\left(-\frac{r^2}{2 \cdot (\sigma_1^2 + \sigma_2^2)}\right)\Big|_{R_c}^{+\infty}. \\
&= \exp\left(-\frac{R_c^2}{2 \cdot (\sigma_1^2 + \sigma_2^2)}\right)
\end{aligned}$$

Therefore, the ICT T_I can be expressed as

$$\{T_I > t\} \equiv \bigcap_{0^+}^{t} E$$

The probability that the two vehicles meet at time t is equivalent to the probability that two vehicles never meet up to time t. Since in our model, the choice of a vehicle at any time t is independent of its previous behavior, E does not depend on t. Let $\eta = \exp\left(-\frac{R_c^2}{2 \cdot (\sigma_1^2 + \sigma_2^2)}\right)$. Thus, we have

$$P_\Xi\{T_I > t\} = P_\Xi\left\{\bigcap_{s=0^+}^{t} E\right\} = \eta^t = e^{-c \cdot t}$$

where $c = \frac{R_c^2}{2 \cdot (\sigma_1^2 + \sigma_2^2)} > 0$. Since $P_\Xi\{T_I > t\} \leq P_\Lambda\{T_I > t\}$, we have $P_\Lambda\{T_I > t\}$ $\leq e^{-ct}$. We drop the index Λ and get

$$P\{T_I > t\} \leq e^{-ct} \tag{3.9}$$

This completes the proof.   ∎

Note that the exponent coefficient c of the tail distribution of the ICT depends on both the communication range R_c and variance σ_1^2 and σ_2^2. Given a communication range, larger variances indicate smaller c. It is not difficult to understand since the larger the variance, the gentler the shape of the PDF of Gaussian distribution of a vehicle's locations will be. This will surely reduce the possibility that two vehicles can "meet". However, since the maximum periodicities are finite, the corresponding σ_1^2 and σ_2^2 are constant, which will not change the nature of the tail distribution function of the ICT from being exponential.

As mentioned before, region boundary can form limitations to the motion of mobile objects under RWP/RWM mobility models [40]. In essence, a traffic influx has a similar effect as a physical boundary to mobile nodes. While boundary restricts the motion of the node in space, a traffic influx puts a limitation in time dimension. Within certain time period, the mobile node will spontaneously run for or passively be "pulled" back towards the traffic influx. Thus, more contact opportunities are added into the system. In real urban scenarios, it is often the case that traffic influxes exist. For example, the underlying road networks by nature take the effect of a huge traffic influx since vehicles have to flow in and out the traffic on roads before getting anywhere. Therefore, the tail distribution of the ICT is at least exponentially fast.

3.3.5 Cross Over from Exponential to Power Law

From the results in Sect. 3.3.4, we can see that the maximum periodicity with regard to the traffic influx plays an essential role in generating the exponential tail distribution of the ICT. As the maximum periodicity increases from a finite value to infinity, the tail distribution of the ICT evolves from exponential decay to power law decay. Then the question is: *How large the maximum periodicity is sufficient to be considered as "infinite"?* Given any large value of the maximum periodicity, as long as our observation lasts long enough, we can always find an exponential tail of the ICT. On the other hand, if the duration of the observation is short, the chances of seeing a vehicle with a large maximum periodicity at the traffic influx are slim. In this case, we are more likely to find a vehicle walking randomly in the region and the consequent power low tail behavior of the ICT.

According to Theorem 3.2, the probability that two vehicles do not meet until time t is

$$P\{T_I > t\} \leq [1 - (1 - \eta)]^t$$

$$= \left[1 - \frac{1}{1/(1 - \eta)}\right]^{1/(1-\eta)\cdot(1-\eta)t} \sim e^{-g(t,\eta)}$$

where $g(t, \eta) \triangleq (1 - \eta)t = \left(1 - \left(1 - \exp\left(-\frac{R_c^2}{2\cdot\left(\sigma_1^2(t)+\sigma_2^2(t)\right)}\right)\right)\right)t.$

If $g(t, \eta) \sim \gamma t$ or $\left(\sigma_1^2(t) + \sigma_2^2(t)\right) \sim \frac{R_c^2}{2\log(1/\gamma)}$, i.e., $\sigma_1^2(t) = (n_1 v_m T_1)^2 = O(1)$ and $\sigma_2^2(t) = (n_2 v_m T_2)^2 = O(1)$, we have $P\{T_I > t\} \sim e^{-\gamma t}$ as expected. In this case, both the maximum periodicity of v_1 and v_2 are finite and does not grow with observation time t. Thus, vehicles are always confined to moving in certain region and the frequency of their meeting is high; If $g(t, \eta) \sim \alpha \log t$, which means $\left(\sigma_1^2(t) + \sigma_2^2(t)\right) \sim \frac{R_c^2}{\log(t/\alpha \log t)}$. Since $T \sim \sigma(t)$, we now consider the interaction among the maximum periodicity T_1, T_2 and the observation time t. There are two cases depending on their expansion scales with time:

(1) *max$\{T_1, T_2\}$ grows much faster than t:* Without loss of generality, we suppose $T_1 = \max\{T_1, T_2\}$. Under this circumstance, T_1 grows much faster than the order of $O\left(\sqrt{\frac{1}{\log(t/\alpha \log t)}}\right)$. Thus, v_1 tends to move more randomly in a broader area. The changes for v_1 to revisit the traffic influx are slim. This definitely prolongs the ICT and the power-law distribution arises.
(2) *Both T_1 and T_2 grow slower than t:* both T_1 and T_2 grow much slower than the order of $O\left(\sqrt{\frac{1}{\log(t/\alpha \log t)}}\right)$. The vehicles can be treated as moving in "bounded" regions under the limited observation time, which implies that they will frequently revisit the traffic influx. This truly introduces an exponential ICT tail distribution.

3.4 Model Evaluation

In this section, we conduct simulations and present the results to support our theoretical conclusion. We first consider our mobility model on an infinitely large region with/without traffic influxes. Moreover, we conduct a set of simulations to demonstrate the impact of underlying road networks to the ICT distribution. In all simulations, the speed of an arbitrary vehicle is chosen randomly from 40 to 80 km/h with a mean value of 60 km/h. The transmission range is set to 100 m.

3.4.1 Impact of Maximum Periodicity

In this experiment, we generate 100 vehicles that run according to our mobility model on a sufficiently large region with one traffic influx at the origin. For the clarity of result analysis, we let all vehicles take equal maximum periodicities. As mentioned before, the interaction between the timescale of the experiment and the timescale of the maximum periodicity of vehicles with respect to the traffic influx is essential in determining the tail distribution of the ICT. To clearly see this interaction, we fix the experiment duration to one month and then increase the maximum periodicity from half of a day to one month to see the possible different types of tail distributions.

Figure 3.6 shows the tail distribution of the ICT between all pairs of vehicles on a linear-log scale for our mobility model. It can be seen that the tail distribution can be approximated by a line in the front part of the curve when the maximum periodicity is set to 12 h and one day, respectively. This indicates an exponential decay on the linear-log scale. As we take larger maximum periodicities, the tail behavior of the ICT starts to evolve from exponential decay to power law decay. This can be better observed from Fig. 3.7 where the scale is log–log. The tail distribution can be approximated by a line when the maximum periodicity equals

Fig. 3.6 Tail distribution of the ICT for the described mobility model under different maximum periodicities: half a day, one day, one week and one month. Figure is drawn on a linear-log scale

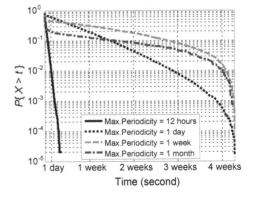

Fig. 3.7 Tail distribution of the ICT for the described mobility model under different maximum periodicities: half a day, one day, one week and one month. Figure is drawn on a log–log scale

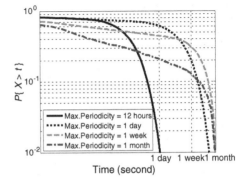

to one month. It is interesting to notice that, at the time scale of one week in this experiment, the tail distribution first exhibits a power law distribution over a time span of about two days from the beginning of time. It then follows an exponential-like decay over a large range of timescale (excluding the fading range caused by limited duration of the observation). Setting the maximum periodicity equal to one week reaches a critical condition where the tail distribution turns from exponential decay into power law decay.

3.4.2 Impact of Real Road Networks

Since urban vehicles always move on roads in the city, the road transportation system can put various constraints to the mobility of these vehicles. To study the impact of the real road networks to the tail distribution of the ICT, we contact a simulation that involves 300 generated vehicles. A vehicle starts to perform random walks on the road networks from the central area. It randomly chooses a distance D and direction each time and runs for three days.

Figure 3.8 shows the destination distributions of all taxies. Since we take a relatively short experiment time and set vehicles to start from the central area. The majority of the vehicles have not reach the boundary of the city. Figure 3.9 shows

Fig. 3.8 The destination and trace distribution on Shanghai map. Every *colored dot* presents the average number of destinations per vehicle per day located in the corresponding 300 × 300 m square area on the map

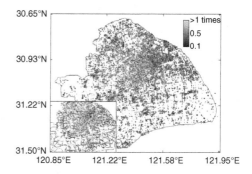

Fig. 3.9 Tail distribution of the ICT for the described mobility model on the real road networks in Shanghai. Figure is drawn on a linear-log scale

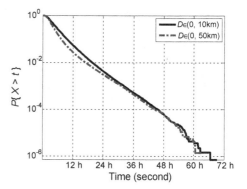

the tail distribution of the ICT with different settings of D. It can be seen that the tail behavior follows an exponential decay even the destinations are randomly chosen. It is because the road networks eventually have the effect of gathering mobile vehicles on roads. This can also be recognized by the inset in Fig. 3.8 where major roads are most visited and have a large amount of traffic.

3.5 Summary

In this chapter, we have demonstrated that the ICTs of taxies have an exponential-like tail distribution by mining the real trace data in Shanghai. To understand the fundamental reason that generates such a tail behavior, we further re-examine the data and have identified the impact of traffic influxes existing in most urban environments by theoretical analysis. We rigorously prove that the tail distribution of the ICT of any two vehicles follows an exponential decay as long as these vehicles have at least one constant traffic influx involved in their normal activities. Our results thus provide fundamental guidelines on design of new vehicular mobility models in urban scenarios, new data forwarding protocols and their performance analysis in VANETs.

With the theoretical and simulation results based on our mobility model, we are highly confident to claim that two vehicles of any type can have the tail distribution of the ICT at least exponentially fast as long as they have one or more common traffic influxes in their normal commuting routines. In real urban environments, there can be more complicated situations. For example, the schedule scheme of traffic lights can have a great deal of impact on vehicular mobility. Two passing vehicles may gain a contact opportunity while waiting for a green light. Note that we do not claim that the traffic influxes in the urban environments are the only factor that can generate the observed exponential tail distribution of the ICT.

There are still many aspects for us to investigate in the future. For example, in more complicated scenarios, there can be multiple traffic influxes. The relationship between the geographical distribution of these traffic influxes and the tail distribution of ICT is uncertain and worth studying. Moreover, it is often assumed in the literature that data transfers can be done instantaneously as soon as two vehicles have a chance to meet. It is definitely not the case in reality where link quality shows very high dynamics. The situation is even worse when consider the same problem in vehicular environments because contacts between vehicles are usually quite short due to high moving speeds and limited communication range. Thus, we will investigate the ultimate end-to-end delay since it can be caused not only by ICTs but also by retransmissions if the data transfer fails in a contact.

Chapter 4
Opportunistic Routing Protocols

4.1 Introduction

In intermittently connected mobile ad hoc networks (MANETs) and delay tolerant networks (DTNs), in order to successfully transfer data from a moving node to another, the node has to first wait until it geographically "meets" other nodes (contact happens) for data-relay. Therefore, data communications are opportunistic in a store-carry-forward fashion, which constantly experiences large end-to-end delays. Data forwarding in VANETs falls in this category. Fast data forwarding which refers to minimizing the end-to-end delay and network traffic at the same time in vehicular networks is the cornerstone of a wide variety of applications. For example, real-time road traffic information can be obtained by exchanging local traffic observations among vehicles.

Provisioning fast data forwarding in vehicular networks, however, is quite challenging due to three reasons. First, even all future contacts are known, finding the fastest path for a given data traffic load is NP-hard [43]. To make things worse, due to the dynamic characteristics of the network, it is very hard to know future communication opportunities between vehicles. With no such information, routing decision made based on past contact statistics can hardly achieve the optimal. Second, with the distributed nature of the network, a vehicle can only have partial information of the network, which make a globally optimal solution is very hard, if not impossible. Last, data communications are via wireless channels and therefore the network resource is restrained, which makes solutions that flood the network [43, 46] infeasible since they introduce prohibitive network traffic.

As communication opportunities are created by the mobility of vehicles, the mobility characteristics of vehicles are central to forwarding algorithms and the ultimate performance in terms of end-to-end delay, delivery ratio and network traffic overhead. Based on the available knowledge about the movement of vehicles, data forwarding algorithms in these networks can be divided into two basic categories: *non-knowledge-based* and *knowledge-based*.

In the non-knowledge-based category, without requiring any information, random walks [47, 48] can be used for data-relay. For a random walk, a vehicle in

H. Zhu and M. Li, *Studies on Urban Vehicular Ad-hoc Networks*,
SpringerBriefs in Computer Science, DOI: 10.1007/978-1-4614-8048-8_4,

the network randomly selects a neighboring vehicle as the next hop to carry a message. Using random walks generates very moderate network traffic but tends to have very large end-to-end delay. By performing multiple walks, both delivery ratio and end-to-end delay can be improved. An extreme case of this is epidemic routing [43, 46], where a message is flooded throughout the network. If there is no bandwidth restriction, using epidemic routing can achieve the minimum end-to-end delay and maximum delivery ratio but generates unacceptable network overhead at the same time. Techniques such as limiting the number of duplicated copies of a message, setting a living timeout for packets and forwarding to selected neighbors can be used to reduce the overhead of epidemic routing.

In the knowledge-based category, there are several methods available to estimate the end-to-end path delay when the future movement of nodes is known ahead of time. For example, S. Jain et al. [49] discuss the path selection algorithms according to how much knowledge about the network topology and network traffic workload being known. The path delay can be calculated as the sum of the expected delay of each hop on this path. A recursive process is deployed in [50] to calculate the minimum end-to-end delivery delay, assuming that the tail distribution of inter-contact times (ICTs, referring to the time durations between two consecutive contacts between the same pair of vehicles) is exponential and ICTs are independent. In reality, however, it is often the case that information about the future movement of vehicles is unavailable. A number of utility-based routing schemes [51, 52] have been proposed for data forwarding based on node history mobility information, such as the contact records, mobility patterns and the rate of connectivity change. In these schemes, a utility function is defined and measured for every other node in the network. If the current message carrier meets a vehicle with a higher utility, the message is forwarded to this vehicle. Based on granularity at which the underlying vehicular mobility is exploited, we divide existing opportunistic forwarding schemes into following two subcategories:

Utilizing microscopic mobility: Algorithms residing in this subcategory extensively investigate pairwise contact properties and their characteristics of nodes to facility data forwarding. For example, in MaxProp [9], likelihood (probability) that a node will encounter the destination of a packet is estimated and used as the forwarding utility. A recursive process has been deployed in [50] to calculate the minimum end-to-end delivery delay, assuming that the tail distribution of ICTs is exponential and ICTs are independent. S. C. Nelson et al. have proposed an encounter-based routing scheme [53] using the rate of encounter of a node as message relay utility. Observing that successive ICTs have strong temporal correlations, Markov chains [54, 69] have been used to predict future contacts. In this category, a utility function is defined and measured for every other node in the network. If the current message carrier meets a node with a higher utility, the message is forwarded to this node. Algorithms in this category would be very effective when delivering packets to those nodes with which a node has prior contact knowledge.

Utilizing macroscopic mobility: In this subcategory, social structures of node mobility are characterized by data forwarding algorithms. Pairwise contacts are aggregated to social graphs that reflect the regular social relationships between nodes. For example, E. M. Daly et al. [55] have proposed a social based routing scheme called SimBet, which assesses similarity and betweenness centrality Packets are routed to most central nodes until a node with higher similarity is met. Then the packet is routing within the community until the destination is reached. P. Hui et al. [56] have proposed a similar social based data forwarding scheme called Bubble Rap, where betweenness centrality is also used to find bridging nodes and communities are explicitly identified by a distributed community detection algorithm. J. Pujol et al. [57] have proposed a forwarding algorithm called FairRoute leveraging two social processes called perceived interaction strength and assortativity to distribute load more evenly among nodes in the network. Recognizing the importance of capturing real social relationships to the performance of data forwarding, T. Hossmann et al. [58] have proposed an online algorithm to infer the optimal aggregation density.

Provisioning fast data forwarding in vehicular networks, however, is quite challenging due to three reasons. First, due to the dynamic characteristics of the network, it is very hard to know future contacts between vehicles. With no such information, routing decision made based on past contact statistics can hardly achieve the optimal. To make things worse, even all future contacts are known, finding the fastest path for a given data traffic load is NP-hard [43]. Second, with the distributed nature of the network, a vehicle can only have partial information of the network, which make a globally optimal solution is very hard, if not impossible. Last, data communications are via wireless channels and therefore the network resource is restrained, which makes solutions like epidemic routing [43, 46] infeasible since they introduce prohibitive network traffic.

In this chapter, we first present a contact-based data forwarding algorithm based on microscopic mobility of vehicles [54]. Specifically, we analyze more than 45 million pairwise contacts resolved from our traces collected in Shanghai and Shenzhen in China to characterize the contact interaction among vehicles. By studying the distribution of ICTs, in addition to the exponential tail distribution, we find that the layout of ICTs also demonstrates an apparent pattern: if a vehicle meets another vehicle at certain time, the probability that the two vehicles meet again at the same time in the following days is very high. With this observation, we characterize the temporal correlation of ICTs and then capture those characteristics with higher order Markov chain models. We then design an opportunistic forwarding algorithm exploiting the temporal dependency of ICTs. In our algorithm, a vehicle estimates the expected delay between a neighboring vehicle and the destination of a message, based on their previous ICTs. If this vehicle has smaller estimation, it forwards the message for data-relay. The goal of our algorithm is twofold: first, we concern the delivery performance in vehicular networks, trying to minimize the end-to-end delay and maximize the delivery ratio; second, since vehicles communicate via wireless channels, we try to minimize the network overhead for data transmission. Through extensive trace-driven simulations, our algorithm can achieve comparable delivery performance as epidemic routing in terms of end-to-end delay and delivery ratio with a very moderate network overhead. Compared with current message forwarding strategy based on the delivery probability or the expected delay, our algorithm can dramatically increase 84 % delivery ratio and reduce 53 % end-to-end delay while generating similar network traffic.

Then, we introduce a social-based routing scheme ZOOM [62] which elegantly manages to capture both microscopic and macroscopic mobility of vehicles in an integrated approach in order to achieve a better tradeoff between end-to-end delay and network traffic cost. The design of ZOOM is based on two key observations found by analyzing the traces. First, consecutive ICTs have strong temporal correlations, which can be utilized to predict future contacts. Second, contact graphs established by aggregating pairwise contacts represent clear social structures. Inspired by these observations, we first train Markov chains to capture the temporal correlations of pairwise contacts, based on which we infer future contact opportunities. We then use centrality to measure the importance of a vehicle in the contact graph. With mobility information in both levels, when two vehicles encounter, the vehicle with shorter expected ICT with the destination will be chosen as the next relay of a packet. If no such information available, the vehicle which has larger network centrality will act as the next data relay. Extensive trace-driven simulation results demonstrate the efficacy of ZOOM design. On average, ZOOM can improve 30 % performance gain comparing to the state-of-art algorithms.

4.2 Contact-Based Routing Protocol Design

Recently, there have been several studies on analyzing mobility characteristics based on empirical trace data collected from urban areas [42, 59] and public transportation systems [44, 60]. These studies mainly focus on the distribution of ICTs, having the observation that vehicles in urban environments tend to meet very frequently. They demonstrate the tail distribution of ICTs can decay exponentially fast. Although the exponential distribution facilitates the problem analysis on the performance bound of routing algorithms, it is not clear *how to design a practical opportunistic forwarding algorithm utilizing the characteristics of ICTs*? We take a data-driven approach in designing and evaluating our opportunistic forwarding algorithm in urban vehicular networks.

4.2.1 Statistics of ICTs

(1) Extraction of ICTs from Trace Data
    Since GPS reports are sent in discrete time, usually on one minute, we use a
    sliding time window to check contacts between a pair of taxies as introduced

in our previous work [44, 45]. Here we make the assumption that two vehicles would have a connection opportunity (called a contact) if their locations reported within a given time window are within the communication range. Although the inaccuracy may be introduced by this assumption and contact extraction algorithm, the essential vehicular mobility characteristics are preserved and therefore the results are very valuable for study.

We refer to *inter-contact time* as the time elapsed between two successive contacts of the same vehicles as defined in [28–30]. Specifically, the inter-contact time is computed at the end of each contact, as the time period between the end of this contact and the beginning of the next contact between the same two vehicles. For example, in Fig. 4.1, inter-contact time d_1 can be computed as the starting time of contact C_2 minus the end time of its previous contact C_1. Table 4.1 shows the statistics of ICTs extracted from three traces.

(2) ICT Distribution Characteristics

We apply the above contact extraction algorithm with a time window of one minute and a communication range of 100 m to each pair of vehicles in all three data sets, respectively (basic statistics are shown in Table 4.1). We plot the tail distribution (CCDF) of inter-contact time over time in linear-log scale in Fig. 4.2. The linear delay of all plots in linear-log scale indicates that the tail distribution of inter-contact time between vehicles drops exponentially. The reason that the ICT tail distribution of vehicles is exponential rather than power law as found in human mobility [28] might be that traffic tends to converge around certain areas in the urban settings, such as the underlying topology of road networks and distribution of residential areas, shopping centers and commercial zones, which enormously increases contact opportunities of vehicles [44, 45]. The exponential distribution implies, to some extent, vehicles meet each other in urban settings very frequently. While exponential distribution is convenient for the problem analysis, we are athirst for the answer to the following question: *how to design a practical opportunistic forwarding algorithm utilizing inter-contact time distribution characteristics*?

To answer the question, it is not enough knowing only the frequency of connection opportunities but particularly the temporal layout or patterns between each inter-contact time within the distribution. Therefore, we examine the probability density function (PDF) of ICTs as shown in Fig. 4.3. It is easy to notice an apparent pattern that the probability reaches local maxima when the length of an inter-contact time equals an integral multiple of one day. This indicates that if a vehicle meets another vehicle at certain time the probability that the two vehicles meet again at the same time in the

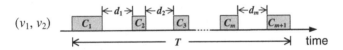

Fig. 4.1 An example of contacts and inter-contact times between a pair of vehicles v_1 and v_2

Table 4.1 Comparison of three data sets

Data set	Shanghai bus	Shanghai taxi	Shenzhen taxi
Number of vehicles	2,501	2,109	8,291
From date	Feb 19, 2007	Feb 1, 2007	Oct 1, 2009
Duration (day)	15	31	31
Granularity (second)	60	15^{*}, 60^{**}	60
Number of contacts	1,229,380	22,053,178	23,968,860
Mean ICT (minute)	31.8	47.6	30.5

* vacant, ** passengers onboard

Fig. 4.2 Tail distribution of the inter-contact time of urban public vehicles in Shanghai (SH) and Shenzhen (SZ)

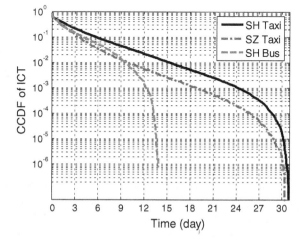

Fig. 4.3 Probability density function of the inter-contact time of the same experimental vehicle sets

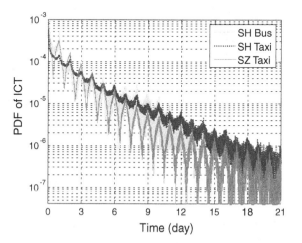

following days is very high. The reason can be explained as follows. Buses can constantly encounter with each other since they have dedicated routes and schedules. Intuitively, taxies behave rather randomly and have higher mobility than buses. Nevertheless, taxi drivers also have their own preferences in choosing serving areas and path planning. Evidence shows that other vehicles of different kinds in urban settings also demonstrate strong mobility patterns during daily routines [63], which constitute regular connection opportunities. In other words, *temporal dependency* of inter-contact time does exist between two vehicles in urban vehicular networks.

4.2.2 Analyzing ICT Temporal Patterns

We examine two specific questions: (1) how historical inter-contact time information is related to the current inter-contact time; and (2) how inter-contact time patterns evolve over time and how much historical information we need to track to capture the inter-contact time patterns over time.

4.2.2.1 Characterizing Temporal Correlations of Successive ICTs

We examine the correlation between ICTs by computing the marginal entropy of ICTs between each pair of vehicles and the conditional entropy of ICTs between a pair of vehicles given their previous M inter-contact times in all of the three data sets.

Although an ICT can be infinitely long in time, due to the fast exponential decay of inter-contact time tail distribution, most inter-contact times are less than a relatively short period of time. For example, in Fig. 4.3, more than 90 % inter-contact times are less than 6 days. Therefore, an inter-contact time T can be specialized into a discrete finite value space as,

$$T' = \begin{cases} \lfloor T/\lambda \rfloor, & if \quad T < T. \\ \lfloor \mathbb{T}/\lambda \rfloor, & otherwise \end{cases} \tag{4.1}$$

where $\mathbb{T}$ is the maximum inter-contact time, and λ is the counting measure. In the rest part of this chapter, without explicit specification, inter-contact times are referred to as their specialized counterparts.

Let X be the random variable representing the inter-contact times between a pair of vehicles. If we have observed N inter-contact times between this pair of vehicles, these inter-contact times can be presented by a vector $T = (t_0, t_1, \ldots, t_{N-1})$ where $t_i \in [0, \lfloor \mathbb{T}/\lambda \rfloor]$, $0 \le i \le N - 1$ denotes the ith inter-contact time. Assume each of these inter-contact times appeared x_j times in T, $0 \le j \le \lfloor \mathbb{T}/\lambda \rfloor$. Thus, the probability of the inter-contact time being j can be computed as x_j/N. Therefore, the entropy of T is:

$$H(X) = \sum_{j=0}^{\lfloor T/\lambda \rfloor} (x_j/N) \log_2 \frac{1}{x_j/N}. \tag{4.2}$$

For $M = 1$ let X' be the random variable for the immediately previous inter-contact time between this pair of vehicles given the inter-contact time X. X' and X have the same distribution when N is large enough. The vector T can be written as $Q = \{(t_i, t_{i+1}) : 0 \le i \le N - 2\}$. Therefore, the joint entropy of X' and X can be computed as:

$$H(X', X) = \sum_{(x', x) \in Q} P(x', x) \log_2 \frac{1}{P(x', x)}, \tag{4.3}$$

where $P(x', x)$ is the number of times (x', x) appearing in Q divided by the total number of elements in Q. With $H(X)$ and $H(X', X)$, the conditional entropy of X given X' is:

$$H(X|X') = H(X', X) - H(X') = H(X', X) - H(X). \tag{4.4}$$

For $M = 2$, let X'' denote the random variable representing the distribution of the previous two ICTs given X. Similarly, the conditional entropy $H(X|X'')$ is:

$$\begin{aligned} H(X|X'') &= H(X'', X) - H(X') \\ &= H(X'', X) - H(X', X), \end{aligned} \tag{4.5}$$

The joint entropy $H(X'', X)$ can be calculated similarly.

Figure 4.4 shows the CDFs of the mean entropy and the mean conditional entropy, for $M = 1$ and 2, over each pair of taxies in the Shanghai data set. It can be seen that the conditional entropy for $M = 1$ is much smaller than the marginal entropy, and that the conditional entropy for $M = 2$ is smaller than that for $M = 1$.

Fig. 4.4 CDFs of marginal entropy and conditional entropy of inter-contact times between each pair of taxies in Shanghai data set

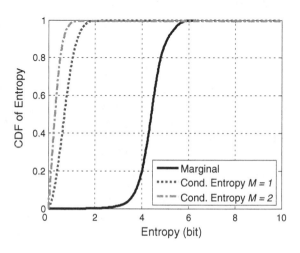

Fig. 4.5 CDFs of marginal entropy and conditional entropy of inter-contact times between each pair of taxies in all data sets

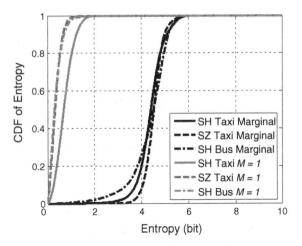

This implies that the uncertainty about the inter-contact time decreases when knowing the previous inter-contact times between the same pair of taxies.

We further examine the entropy and conditional entropy for vehicles in all data sets. Figure 4.5 shows the results for marginal entropy and conditional entropy when $M = 1$. It is clear to see that the conditional entropy is much smaller than the marginal entropy for all types of vehicles. In addition, all entropy distributions are very close. Buses have much smaller conditional entropy than taxies in Shanghai. Therefore, although a pair of buses can have as many inter-contact times as a pair of taxies do, the inter-contact times between buses are more correlated than those between taxies. Interestingly, taxies in Shenzhen also have much smaller conditional entropy than taxies in Shanghai. This suggests that taxies in Shanghai operate more randomly with less interference of drivers than taxies in Shenzhen.

4.2.2.2 Evolution of ICT Patterns

In order to establish informed message forwarding strategy utilizing inter-contact time temporal patterns, we divide time into short time slots and examine the correlation between the distribution of inter-contact times between a pair of vehicles in time slot t and that in time slot $t-n$, increasing n from one to a large number. We use *redundancy* to quantify the correlation. Specifically, the inter-contact times between this pair of vehicles in time slot t forms a time series $T_t = (n_0, n_1, \ldots, n_{|t|-1})$, where $|t|$ is the length of a time slot and n_i is the number of inter-contact times occurred at time $i(0 \leq i \leq |t| - 1)$. We also have the time series of inter-contact times in time slot $t - n$, T_{t-n}. We compute the mutual information of T_t and T_{t-n}, $I(T_t, T_{t-n})$ via the joint entropy $H(T_t, T_{t-n})$ and the marginal entropy $H(T_t)$ and $H(T_{t-n})$ as follows:

$$I(T_t, T_{t-n}) = H(T_t) + H(T_{t-n}) - H(T_t, T_{t-n}). \tag{4.6}$$

We define the redundancy of X_{r1} and X_{r2} by

$$R(T_t, T_{t-n}) = \frac{I(T_t, T_{t-n})}{H(T_t) + H(T_{t-n})}. \tag{4.7}$$

We compute the mean redundancy averaged over all pairs of vehicles in Shanghai bus data set from March 5, Shanghai taxi data set from March 3 and Shenzhen taxi data set from October 31, respectively. Time is divided into time slots of 4 h. Figure 4.6 shows the result for $n = 1$ to 84 (a period of 2 weeks). It can be seen that the layout of inter-contact times in a period of time has higher correlation with historical information when the time difference is a multiple of one day for all types of vehicles. Buses have higher redundancy than taxies. Therefore, the inter-contact times between buses are more predictable. Interestingly, the redundancy with Shanghai taxies achieves higher values on even numbers of days than on odd ones whereas the redundancy with Shenzhen taxies is more homogeneous throughout the whole period of time, having larger variances. This should reflect the different shift rules of taxies in these two cities. In Shanghai, taxi drivers usually shift every 24 h so a taxi behaves very differently on every day but very similarly on every other day. The case in Shenzhen, where drivers shift twice a day (e.g., 7 am and 5 pm), is that a taxi behaves differently during the daytime but similarly on every day.

To better understand how much history data should be considered in capturing the inter-contact time patterns, we examine the redundancy between the layout of inter-contact time in time slot t and the aggregated historical information from $t - 1$ to $t - n$, i.e., $\sum_{i=1}^{n} T_{t-i}$. We plot the average redundancy over all pairs of vehicles in the three data sets shown in Fig. 4.7. It is clear that the redundancy increases until n reaches to about 3 weeks. This implies that information older than 3 weeks does not help in capturing inter-contact time temporal patterns.

Fig. 4.6 Mean redundancy of the layout of inter-contact times between two different time slots over all pairs of vehicles in the three data sets

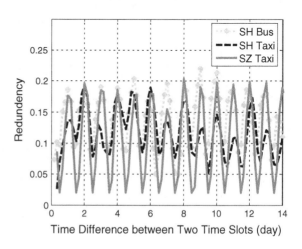

Fig. 4.7 Mean redundancy of the layout of inter-contact times with aggregated history ICTs over all pairs of vehicles in the three data sets

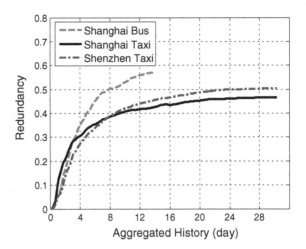

4.2.3 Opportunistic Forwarding Algorithm Design

The analysis based on empirical vehicular trace data in above suggests that it is possible to predict when the next connection opportunity between a pair of vehicles will probably occur based on their recent inter-contact times. This enlightens the design of new opportunistic forwarding algorithms [54] in urban vehicular networks. In this section, we first capture the inter-contact time temporal patterns between each pair of vehicles using higher order Markov chain models. Then, we describe our opportunistic forwarding strategy and discuss the algorithm parameter configuration in terms of system performance and memory cost.

4.2.3.1 Markov Chain Model of k-th order

The class of finite-state Markov processes (Markov chain models) is rich enough to capture a large variety of temporal dependencies. In Markov chain models, the current state of the process depends only on a certain number of previous values of the process, which is the order of the process. By (1), continuous values of inter-contact times can be specialized into finite state space, $\mathcal{S} = \{0, 1, \ldots, \lfloor \mathbb{T}/\lambda \rfloor\}$. Thus, we can establish a k-th order Markov chain to represent the temporal dependency of inter-contact time between a pair of vehicles. The number of states is $(\lfloor \mathbb{T}/\lambda \rfloor + 1)^k$ and the number of conditional probabilities is $(\lfloor \mathbb{T}/\lambda \rfloor + 2)^k$.

More specifically, let $\{x_i\}_{i=1}^{n}$ be an observed sequence of inter-contact times between this pair of vehicles. The k-order state transition probabilities of the Markov chain can be estimated for all $a \in \mathcal{S}$ and $\underline{b} \in \mathcal{S}^k$, $\underline{b} = (b_1, b_2, \ldots, b_k)$ as follows. Let $n_{\underline{b}a}$ be the number of times that state $\underline{b}$ is followed by value a in the sample sequence. Let $n_{\underline{b}}$ be the number of times that state $\underline{b}$ is seen and let $p_{\underline{b};a}$

denote the estimation of the state transition probability from state $\underline{b}$ to state $(b_2, \ldots, b_k, a)$. The maximum likelihood estimators of the state transition probabilities of the k-th order Markov chain are

$$
p_{\underline{b};a} = \begin{cases} n_{\underline{b}a} / n_{\underline{b}}, & \text{if} \quad n_{\underline{b}} > 0 \\ 0, & \text{otherwise} \end{cases} \tag{4.8}
$$

4.2.3.2 Opportunistic Forwarding Strategy

In order to acquire the knowledge of inter-contact patterns, a vehicle first collects recent inter-contact times between itself and all other vehicles. Meantime, it establishes a k-th order Markov chain for each interested vehicle in the network by determining the state transition probabilities according to (9). As a new inter-contact time comes, the vehicle also updates the corresponding Markov chain. It then uses the established Markov chain model as guidance to conduct future message forwarding. Specifically, when a vehicle v_1 encounters vehicle v_2, v_1 examines all messages stored in the buffer of v_2. Suppose v_d is the destination of such a message. Let $\underline{b}_{v_1, v_d}$ denote the current state in the k-th order Markov chain between v_1 and v_d. The estimated delay of the next contact between v_1 and v_d, $\varepsilon_{delay}^{v_1, v_d}$ can be calculated as,

$$
\varepsilon_{delay}^{v_1, v_d} = \sum_{a=0}^{\lfloor T/\lambda \rfloor} p_{\underline{b}_{v_1, v_d};a} \cdot a. \tag{4.9}
$$

Vehicle v_1 will act as the next relay for this message if one of the two following cases happens: 1) v_1 is the destination of this message, i.e., $v_1 = v_d$, and 2) v_1 is a better candidate for relaying this message if the estimated delay of the next contact between v_1 and v_d is shorter than that between v_2 and v_d, i.e., $\varepsilon_{delay}^{v_1, v_d} < \varepsilon_{delay}^{v_2, v_d}$. After transmitting the message to v_1, v_2 simply removes this message from its buffer. Similarly, v_2 will also check messages carried by v_1 and relay messages if needed.

4.2.3.3 Algorithm Parameter Configuration

In our opportunistic forwarding algorithm, there are four key parameters that impact the system performance, namely the maximum inter-contact time in consideration $\mathbb{T}$, the counting measure λ, the order of Markov chain models k and the length for learning stage. In addition, vehicles can have large but limited memory.

Given $\mathbb{T}$, a small counting measure λ will increase the number of states in the Markov chain models, preserving more detailed information at a price of larger memory consumption. On the other hand, if λ equals $\mathbb{T}$, there is only two states in the Markov chain. Thus, a pair of vehicles can only judge the probability that the

Fig. 4.8 The memory cost versus counting measure λ and the order of Markov models k

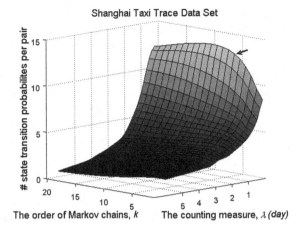

delay of their next connection is larger than $\mathbb{T}$. This has less sense in helping message forwarding. Intuitively, with more detailed information, vehicles can predict more accurately about next communication opportunities. Therefore, there is a tradeoff between memory cost and system performance. Given the state space of a Markov chain model, simply increasing k will not help increase the number of state transition probabilities. The order of Markov chain models k can be determined by conducting AIC tests [20]. Due to the limitation of space, we omit the details.

Figure 4.8 shows an example of the average number of state transition probabilities per pair of vehicles in Shanghai taxi trace data set. It can be seen that the number of state transition probabilities reaches the maximum when λ takes the minimum value (i.e., 4 h in this example) and k equals six.

From the analysis in Sect. 4.2.2, it is clear that increasing the length of learning stage will definitely help improving the accuracy of estimation for next connections. It also suggest that history information that is old than about 3 weeks will not help. Note that all Markov chains are established along with the movement of vehicles in real time. The performance of the proposed opportunistic forwarding algorithm will gradually improve as more learning data becomes available. We will further examine the effect of λ, k and the length of learning stage through trace-driven simulations in the next section.

4.2.4 Performance Evaluation

4.2.4.1 Methodology

In this section, we compare our opportunistic forwarding algorithm with several alternative schemes:

- **Epidemic**. In this scheme [46, 47], vehicles exchange every packet whenever they experience a contact. If vehicles have infinite buffer size, using epidemic routing will find the shortest path between the source and destination vehicles and therefore has the shortest end-to-end delay. On the other hand, since there is no control on data forwarding, it also generates a tremendously large volume of network traffic, overwhelming limited wireless bandwidth.
- **Minimum Expected Delay (MED)**. This scheme [49] utilizes the expected delay metric to guide data forwarding. Expected delay is used to estimate the expected delay between two vehicles v1 and v2 based on contact records. Given the contact record shown in Fig. 4.1, expected delay can be calculated as $D(v_1, v_2) = \frac{\sum_{i=1}^{m} d_i^2}{2T}$. When conducting packet forwarding, the vehicle with the minimum expected delay is chosen as the next hop.
- **Maximum Delivery Probability (MDP)**. This scheme [9, 61] utilizes the delivery probability metric to guide data forwarding. Delivery probability is designed to reflect the contact frequency, i.e., how often two vehicles meet each other. For example, if the contact record between vehicles v1 and v2 is shown in Fig. 4.1, the delivery probability between vehicles v1 and v2 can be calculated as $P(v_1, v_2) = 1 - \frac{\sum_{i=1}^{m} d_i}{T}$. Upon selecting a next-hop vehicle to forward a packet, a vehicle prefers the neighbor with the maximum delivery probability.

We consider three important metrics to evaluate the performance of our algorithm and the above schemes:

(1) *Delivery ratio*. It refers to the success ratio of the number of successfully delivered packets to the total number of packets at the end of an experiment of certain time.
(2) *End-to-end delay*. It refers to the delay for a packet to be received to its destination. It can be calculated by accumulating every delay of each hop. We only calculate end-to-end delay for successfully delivered packets.
(3) *Network traffic per packet*. It refers to the average network cost per packet, calculated by dividing the total number of data forwarding by the number of packets.

In the following simulations, we evaluate the performance of our opportunistic forwarding algorithm in terms of the above metrics, using real trace data from Shanghai taxis, Shenzhen taxis and Shanghai buses. From each data set, we randomly choose 500 vehicles. We then extract contact records between each pair of vehicles for all selected vehicles, using the algorithm described in Sect. 4.2.3. We use the contact records in the first 3 weeks (one week for bus due to the limited available data) as the learning stage for all alternative schemes and use the last week for data transmission. At beginning of each experiment, we inject 100 packets using a Poisson packet generator with a mean interval of 10 s. For each packet, the source and destination are randomly chosen among all vehicles in each data set. Here we assume that two vehicles can always successfully conduct all data transmission when they have a contact.

4.2.4.2   Effect of Algorithm Parameters

We first examine the effects of protocol parameters to the network delivery performance. The maximum inter-contact time $\mathbb{T}$ is set to be 6 days (90 % confidence interval). We vary the counting measure λ from 4 h to 6 days at an interval of 4 h and vary the order of Markov chain k from one to 20 at an interval of one. For each value of λ and k, we run the experiment 50 times and measure the average results.

Figure 4.9 shows the end-to-end delay based on Shanghai taxi. The minimum end-to-end delay can be achieved with the smallest λ equal to four hours and k equal to six in this case. It is clear that increasing λ will result in larger end-to-end delay. To some extent, increasing k will not reduce the end-to-end delay. Figure 4.10 shows the delivery ratio as a function of λ and k. It can be seen the delivery ratio reaches the maximum with the smallest λ and k equal to six. The delivery ratio increases very fast as k increases in the beginning but after that it starts to decrease gradually. When λ varies from 4 h to 6 days, the delivery ratio decreases. These results verify the conclusion described in Sect. 4.2.3.3. We also check the effect of the configuration of λ and k to the delivery performance on Shanghai buses and Shenzhen Taxies. The result is similar, i.e., taking the smallest λ will get the best performance with k equal to five based on Shanghai bus data and six based on Shenzhen Taxi data.

4.2.4.3   Effect of Learning Stage

In this simulation scenario, we examine how much history information is essential for setting up our models. We apply a small λ and the corresponding optimal

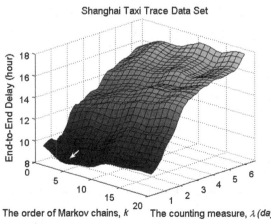

Fig. 4.9 The end-to-end delay versus counting measure λ and the order of Markov models k

Fig. 4.10 The delivery ratio versus counting measure λ and the order of Markov models k

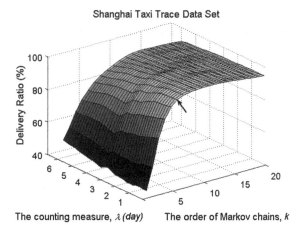

Fig. 4.11 The end-to-end delay versus the length of learning stage based on Shanghai taxi data

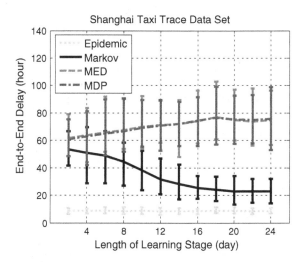

configuration of k and gradually increase the time for learning. For example, in Shanghai taxi trace data, we set $\lambda = 30$ min and $k = 6$ and use the trace in last week, from Feb 25 to Mar. 3, for data transmission. We increase the time for learning from one day (i.e., Feb. 24), 2 days (i.e., Feb. 23, 24) till 24 days (i.e., Feb. 1–24). For each training time, we run the experiment 50 times and measure the average results.

Figure 4.11 shows the end-to-end delay as the length of learning stage grows. It can be seen that, with more history information available, our algorithm can dramatically reduce the average end-to-end delay from 53.62 to 22.87 h. When the length of learning stage is larger than 19 days, the end-to-end delay hits a plateau and stabilizes. This is consistent with our observation in Sect. 4.2.2.2 that history information older than 3 weeks will not contribute more. Surprisingly, as the learning time grows, both MED and the MDP schemes have larger end-to-end

Fig. 4.12 The delivery ratio versus the length of learning stage based on Shanghai taxi data

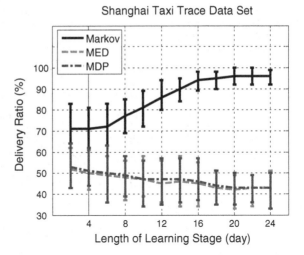

delay. Since these schemes are based on aggregated characteristics of inter-contact times, they cannot fully utilize the temporal dependency of vehicular mobility. The MED and the MDP schemes achieve the minimum end-to-end delay of 61.62 and 61.02 h, respectively, using one day for learning. The epidemic scheme has the minimum end-to-end delay of 8.6 h.

We plot the delivery ratio as a function of learning time shown in Fig. 4.12. We omit results from the epidemic scheme since it can always get a 100 % delivery ratio in this setting. The Markov scheme can reach to a 96 % delivery ratio when the length of learning stage is larger than 3 weeks. It can also delivery about 84 % more packets, compared with the best performance of the MDP and the MED schemes (53 and 52 %). Figure 4.13 shows the average network traffic per packet generated in the network. It can be seen that it takes three more hops on average to deliver a packet using the Markov scheme than using the MED and the MDP schemes to achieve best performance. The epidemic scheme has the largest network cost of 1.87×10^5 hops. In summary, our scheme can achieve comparable delivery performance as the epidemic scheme with a conservative network cost. We also examine the effect of learning stage to the network performance based on Shenzhen taxi data and Shanghai bus data. Results are presented in Table 4.2.

4.2.4.4 Effect of Multiple Paths

In previous simulations, each packet follows only one path, i.e., at any time, at most one copy of a packet exists in the network. In this simulation, we extend our algorithm to allow multiple copies of a packet, thus to improve delivery performance in terms of shorter delay and higher delivery ratio. We consider two multiple path forwarding strategies:

Fig. 4.13 The network
traffic per packet versus the
length of learning stage based
on Shanghai taxi data

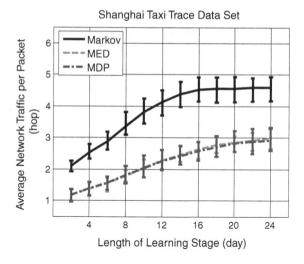

Table 4.2 Performance comparison of all schemes

Shenzhen taxies	Min. end-to-end delay (h)	Max. delivery ratio (%)	Network traffic (hop)	Shanghai buses	Min. end-to-end delay (h)	Max. delivery ratio (%)	Network traffic (hop)
Markov	23.68	83	3.34	Markov	34.12	95	2.33
MED	49.70	40	1.82	MED	74.90	53	1.47
MDP	48.81	41	2.04	MDP	74.29	53	1.47
Epidemic	3.34	100	1.25×10^5	Epidemic	11.67	100	2.06×10^5

(1) **Better Candidate**. In this strategy, instead of removing a packet from its buffer after message forwarding, a vehicle keeps a copy of a packet and can always forward the packet to other candidate vehicles in the future;

(2) **Ever-best Candidate**. In this strategy, a vehicle also keeps a copy of a packet but only transmits the packet to a candidate that has the ever-best delay estimation among all previous candidates it has chosen.

We apply these two strategies to our Markov scheme, the MDP and the MED schemes, and conduct experiments with the same configuration as that in the above simulation except all available data in learning stage are used. The end-to-end delay, delivery ratio and the network traffic per packet based on Shanghai taxi data are shown in Figs. 4.14, 4.15 and 4.16, respectively. It can be seen that the proposed scheme can achieve appealing delivery performance (22.87-hour end-to-end delay and 96 % delivery ratio) even with one-path forwarding. By conducting multiple path forwarding, the MED and MDP schemes can achieve smaller end-to-end delay and larger deliver ratio but at a very high network cost.

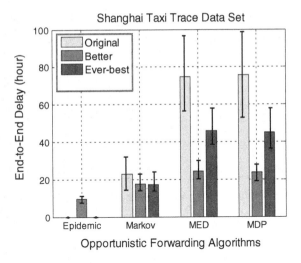

Fig. 4.14 The end-to-end delay versus opportunistic forwarding algorithms based on Shanghai taxi data

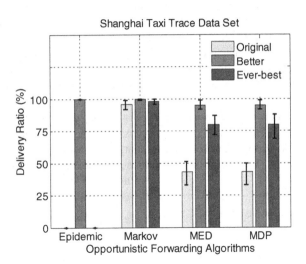

Fig. 4.15 The delivery ratio versus opportunistic forwarding algorithms based on Shanghai taxi data

4.3 Social-Based Routing Protocols

By collecting and analyzing contacts between two vehicles, contact-based data forwarding algorithms try to find predictive statistics about this pair of vehicles, such as the frequency and the spatial-temporal distributions of contacts and ICTs and then use such information to guide data forwarding. In general, contact-based algorithms can be very efficient when forwarding packets between regularly encountered nodes but less effective when no prior contact knowledge is available.

Fig. 4.16 The network traffic per packet versus opportunistic forwarding algorithms

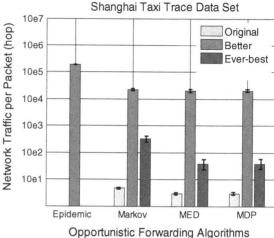

Recently, social network analysis has been proposed as a general and powerful tool to forward data in DTNs. By *aggregating* past observed pair wise contacts into a *social* graph, data forwarding algorithms focusing on *social-level mobility* study the network structure, which push packets towards nodes with more important network positions. In general, social-based algorithms leverage the knowledge of network structures to route packets. The rationale of algorithms in this category is based on the "small world" [64] phenomenon in social networks. However, short chains of acquaintances do not mean that the delay for delivering messages between a pair of nodes is necessarily short. Furthermore, it is less efficient when forwarding data between regularly-met vehicles.

In this section, we introduce an innovative data forwarding algorithm, called ZOOM, which elegantly leverage both contact-level mobility and social-level mobility to tackle those challenges in solving the fast data forwarding problem in vehicular networks.

4.3.1 Macroscope Mobility of Social Relationship

Contacts of vehicles actually reflect the complicated social activities of human beings, the characteristic macroscopic structures of human relationships may create complex patterns of contacts, which cannot easily be observed or well understood by only analyzing individual pairwise contacts. For example, people meet "strangers" by chance, "friends" by intention or "familiar strangers" because of their similarity of mobility patterns. In this section, we examine the vehicular mobility from a more macroscope perspective.

4.3.1.1 Establishing Contact Graph

We establish a static and weighted contact graph $\mathbb{G}(N, E)$ for each trace by aggregating the entire sequence of contacts between a pair of vehicles. Each vehicle i is a node of the graph, $n_i \in N$, and the edge $e_{ij} \in E$ represents node i and j have certain acquaintance between them. The key to establishing a meaningful contact graph is the metric used to aggregate contacts, which determines whether two nodes share a link and the strength of this connection if exists. Various metrics, such as the number of total contacts observed [56], the age of last contact [65], and the contact frequency and total duration [56], have been used to derive edge strengths. In our study, we use a sliding window to consecutively check the ratio of time with contacts observed to the total period of a trace, called *contact ratio*. There is an edge between two nodes in the contact graph if the contact ratio is higher than a threshold and the weight on this edge takes the contact frequency value. The main reason that we use this metric to aggregate contacts is to reduce the influence of random (*unexpected*) contacts in vehicular networks and comprise as many "regular" relationships as possible. Figure 4.17 illustrates the contact graph established on Shanghai taxi trace with sliding window size of one day and contact ratio equal to 60 %.

4.3.1.2 Revealing Social Structures

We study the social properties of the contact graph of each trace and examine the degree distributions. The degree of a node in the contact graph is the number of

Fig. 4.17 Contact Graph of Shanghai Taxi Trace containing 1,226 nodes, which is highly structured with 56 communities

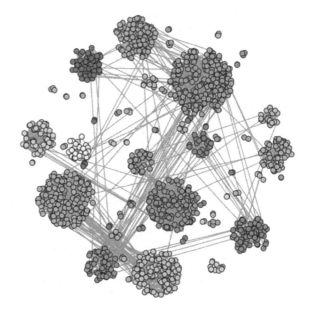

Fig. 4.18 The CCDF of the vehicle degree on all traces under semi-logarithmic scale

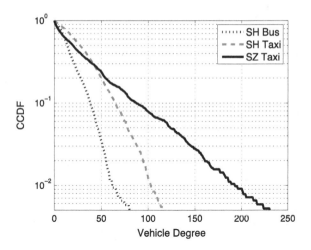

edges incident on this node. We define p_k to be the fraction of nodes in the contact graph that have degree k and plot the complementary cumulative distribution function (CCDF) $P_k = \sum_{k'=k}^{\infty} p_{k'}$. Figure 4.18 shows the CCDF of vehicle degree on all traces under semi-logarithmic scale. It is clear to see that all degree distributions have exponential tails,

$$P_k = \sum_{k'=k}^{\infty} p_{k'} \sim \sum_{k'=k}^{\infty} e^{-\frac{k'}{\alpha}} \sim e^{-\frac{k}{\alpha}}. \tag{4.10}$$

Similar degree distributions have been seen with different networks such as the power grid and railway networks [66]. In contrast, random graphs, where each edge is present or absent with equal probability, have binomial (Poisson in the limit of large graph size) degree distributions.

We further check whether there are *communities* embedded in a contact graph. A community is defined as a subset of nodes with stronger connections between them than towards other nodes, which generally implies a social group. The *modularity* [67] can be used to evaluate the partition of nodes to communities, which is defined as

$$Q = \frac{1}{2m} \sum_{ij} \sum_{r} \left(A_{ij} - \frac{k_i k_j}{2m} \right) S_{ir} S_{jr}, \tag{4.11}$$

where m is the total number of edges, A_{ij} is the element of the adjacency matrix (if there is an edge between node i and j, $A_{ij} = 1$; otherwise, $A_{ij} = 0$), k_i and k_j are the degree of node i and j, respectively, and $S_{ir} = 1$ if node i belongs to group r and zero otherwise. Finding the optimal community structure for a contact graph in terms of maximal modularity is an NP-complete problem. We use the Louvain algorithm [68] which iteratively moves each node to an existing community and merges two communities if doing so can maximize the modularity. We choose this algorithm because it has been reported to be fast and has good or better community partition comparing with other algorithms on a different number of graphs [68].

The modularity and the number of found communities for all traces are listed in Table 4.3. From the list, we have the following observations: (1) the modularity values vary in traces but overall are quite high. This implies that urban vehicular networks are highly structured rather than randomly connected $(Q = 0)$. High modularity $(Q > 0.3)$ can also be seen in other social and biological networks [67]. (2) Buses have higher modularity than taxies. This is easy to understand since buses have dedicated routes and schedules, which makes contacts constant and stable.

4.3.2 Impact of Mobility on Routing Algorithms

In the store-carry-and-forward scenario, the performance of a particular opportunistic forwarding algorithm heavily relies on its capability to accurately capture the underlying mobility of vehicles. In this section, we discuss the impact of different mobility scales to the performance of routing algorithms.

4.3.2.1 Algorithms Utilizing Contact-Level Mobility

By collecting and analyzing contacts between two vehicles, it is possible to obtain detailed knowledge about this pair of vehicles such as the contact frequency [9] and the expected delay [54, 69]. Such local knowledge can be used to determine data-relays for a routing algorithm.

To illustrate the performance of algorithms utilizing contact-level mobility of vehicles, we examine a greedy algorithm, called *Future*, in which all future contacts between vehicles are known. In *Future*, a vehicle with messages always chooses a neighboring vehicle which has the shortest delay with the destination. We randomly select 1,000 pair of vehicles as the source and destination of 1,000 messages, using Shanghai taxi trace. Figure 4.19 shows the CDF of end-to-end delay over all messages using *Future* and Epidemic routing [46, 47], where a vehicle always forwards its messages to any vehicle it meets. It can be seen that *Future* performs well but experiences larger end-to-end delay comparing to

Table 4.3 Comparison of there data sets

Data set	Shanghai bus	Shanghai taxi	Shenzhen taxi
Number of vehicles	2,501	2,109	8,291
Duration (day)	15	31	31
Number of contacts	1,229,380	22,053,178	23,968,860
#Communities	29	56	43
Q	0.8733	0.8471	0.6230

Fig. 4.19 The CDF of end-to-end delay over 1,000 random generated messages using different algorithms

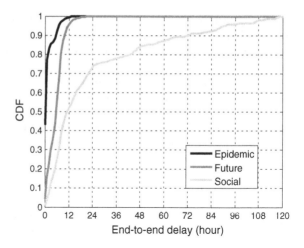

Epidemic routing. For example, above 90 % messages can be delivered within 6 h using Epidemic routing whereas Future can only reach about 60 %.

The main reason for *Future* being sub-optimal is that, without global contact information of other vehicles, *Future* may only find local optimal routing path. Moreover, most vehicles, due to limited mobility, only have contacts with a small portion of other vehicles. For example, Fig. 4.20 plots the CDF of the ratio of the number of vehicles met by a vehicle to the total number of vehicles in all traces. It can be seen that most Shanghai taxis can only "see" 10 % of all taxis. For Shanghai buses, the proportion reduces to about 5 % due to the limitation of fixed itineraries and schedules of buses. Comparing to Shanghai taxis, Shenzhen taxis have higher proportion of encountered taxis. The reason seems to be that Shanghai City has larger area than Shenzhen City (three times bigger). Given the

Fig. 4.20 The CDF of the number of encountered vehicles on all datasets

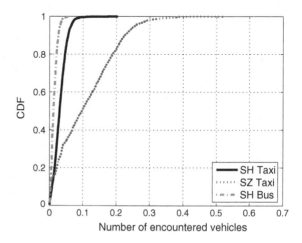

same mobility of a taxi, Shenzhen taxies have more opportunities to meet other taxies. Nevertheless, the proportion is still low (e.g., 80 % taxies only "see" 20 % other taxies).

The consequence of limited view about the whole network is that when a vehicle v_1 is requested to deliver a message to vehicle v_d, it is very likely that v_1 has no knowledge about v_d. To make things worse, when v_1 encounters another vehicle v_2, very likely, v_2 has no information about v_d either. In that case, v_1 has to carry the message until it meets v_d or another vehicle which knows v_d. This will increases the end-to-end delay.

In summary, data forwarding algorithms based on contact-level mobility are effective when delivering packets among "familiar" vehicles with prior contact knowledge but less efficient for "stranger" vehicles.

4.3.2.2 Algorithms Utilizing Social-Level Mobility

With contact graph and the social structure observed in the contact graph as described in Sect. 4.3.1, a data forwarding algorithm can utilize the social features of nodes or the network to facilitate data forwarding. For example, a greedy hill-climbing procedure is conducted in the network, seeking for more "central" or "popular" nodes in the graph using social network analysis metrics (e.g., centrality and similarity) as data carriers [55, 56]. The rationale of such data forwarding algorithms is based on the "small world" phenomenon in social networks which comes from the observation that individuals are often linked by a short chain of acquaintances (e.g., "six degrees of separation" [64]).

In vehicular network scenario, however, the process of seeking for central nodes as data-relays does not match the goal of the fast opportunistic forwarding problem. A hop in the short paths in social networks may actually undergo a tremendous delay, which is prohibitive for fast data forwarding. In the extreme case, a vehicle can hold a message until it finally meets the destination of the message, which is optimal in terms of the number of hops required to forward the message but definitely not the optimal for minimizing the end-to-end delay. In order to verify our argument, we also conduct an experiment using the same setting as the one described in the above subsection. We evaluate the Sim-Bet algorithm [55] with each node knowing its global social betweenness and similarity values in the contact graph established from Shanghai taxi trace. In SimBet algorithm, a neighboring vehicle is selected as the next relay if a weighted betweenness and similarity utility regard to the destination increases. The CDF of end-to-end delay is shown in Figure 4.19. We find that the overall end-to-end delay is quite large. For example, it requires almost 3 days for 90 % messages to be delivered.

The main reason for algorithms utilizing social-level mobility characteristics experiencing large delay is that the social features of vehicles and the network is based on long-term statistics of contacts, which discards the short-term dynamics happening between each pair of vehicles. Specifically, due to the high mobility of vehicles, contacts between two vehicles evolve fast which makes the contact aggregation hard to be accurate. For example, we divide a day into four time slots of 6 h and distribute all contacts into respective time slots according to the time when a contact happened. Let X be the random variable that represents the number of contacts exists in a time slot. We then plot the CCDF of the ratio of the standard deviation $\sigma(X)$ to the mean value $E(X)$ for all pairs of vehicles in each trace in Fig. 4.21. It can be seen that contacts occur quite uneven during a day. For example, for Shenzhen taxies, the probability that the number of contacts between a pair of vehicles can vary 60 % comparing to their average number of contacts in a day is above 80 %.

Furthermore, without specific contact-level mobility characteristics, social-based algorithms perform less efficient when routing messages among vehicles with acquaintances. For cxample in Fig. 4.22, suppose that v_1 has a packet for v_d and encounters v_2 and v_3 at the same time. If v_3 is more "central" than v_2 in the network, v_1 will forward the packet to v_3 even if v_2 will meet v_d sooner than v_3 (i.e., $t_1 < t_2$).

Fig. 4.21 The CCDF of the ratio of standard deviation to the average of contact distribution in a discrete time slot

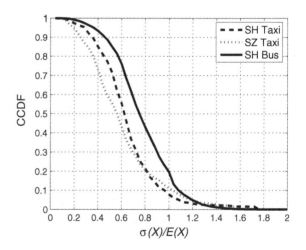

Fig. 4.22 Opportunistic data forwarding scenario, where dashed arrow lines denote the trajectories of vehicles and the disk shades denote the wireless communication range

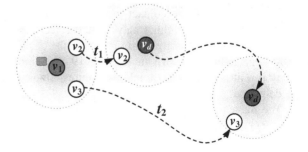

In summary, algorithms based on social-level mobility are effective especially when delivering packets to those "stranger" vehicles but less effective due to the lack of detailed contact-level mobility information.

4.3.3 Design of ZOOM

4.3.3.1 Design Overview

From the analysis above, an ideal opportunistic forwarding algorithm should take both contact-level and social-level mobility into account. To this end, we design an innovative opportunistic data forwarding algorithm, ZOOM, which elegantly manages to capture two levels of vehicular mobility in an integrated approach. The core idea of ZOOM is for each vehicle to locally maintain a list of recent contacts with each other encountered vehicle. With the list of past contacts, a vehicle first trains a k-order Markov chain for each other vehicle which can be used to predict the next contact with that vehicle. In addition, it also assesses its position in the network using ego betweenness centrality based on its ego contact graph aggregated from all its contact lists. With the knowledge of the predicted future contact and the ego betweenness, when two vehicles meet, a vehicle carrying a packet first compares its predicted contact delay with the destination of the packet with that of the other vehicle. The vehicle with shorter contact delay estimation will act as the next data-relay. If both vehicles have no contact predictions with the destination, the vehicle having more important position in the network is chosen to carry the packet.

In the following subsections, we first describe our method to capture the contact-level mobility using k-order Markov models to predict future contacts. Then we present the techniques to establish social-level mobility by aggregating fast evolving contacts and calculating the network position of vehicles using ego betweenness. Finally, we describe the opportunistic forwarding strategy of ZOOM.

4.3.3.2 Predicting Contact-Level Mobility

With the strong temporal correlations of successive ICTs embedded in vehicular mobility as described in Section III, we predict ICTs using Markov chains of k-th order [54].

More specifically, let $\{x_i\}_{i=1}^n$ be an observed sequence of ICTs between this pair of vehicles. The k-order state transition probabilities of the Markov chain can be estimated for all $a \in \mathcal{S}$ and $\underline{b} \in \mathcal{S}^k$, $\underline{b} = (b_1, b_2, \ldots, b_k)$ as follows. Let n_{ba} be the number of times that state $\underline{b}$ is followed by value a in the sample sequence. Let n_b be the number of times that state $\underline{b}$ is seen and let $p_{b;a}$ denote the estimation of the state transition probability from state $\underline{b}$ to state $(b_2, \ldots, b_k, a)$. The maximum likelihood estimators of the state transition probabilities of the k-th order Markov chain are

$$
P_{\underline{b};a} = \begin{cases} n_{\underline{b}a} / n_{\underline{b}}, & \text{if} \quad n_{\underline{b}} > 0. \\ 0, & \text{otherwise} \end{cases} \tag{4.12}
$$

4.3.3.3 Establishing Social-Level Mobility

To utilize social-level mobility to facilitate opportunistic data forwarding in vehicle networks, ZOOM has to deal with two challenges. One is to accurately aggregate high-dynamic contacts so that the real social-level mobility can be reflected in the contact graph. The other one is to accurately assess the importance of an individual vehicle in the network without the global information.

(1) Aggregating Evolving Contacts
   We use the aggregation method introduced in Subsection III C. We use 6 h as the best sliding time window for contact aggregation as it is reported in [54] that the redundancy of ICTs reaches the minimum when the time difference between two small sets of consecutive ICTs equals to 6 h. This also implies that the maximum mobility diversity can be observed within 6 h. Increasing the size of the sliding window will reduce the mobility diversity which degrades the accuracy of contact aggregation. It is important to note that different and more sophisticated aggregation schemes are possible, such as online algorithms [58]. Our goal here is to demonstrate that capturing social-level mobility as a complementary counterpart of contact-level priors can significantly improve the performance of opportunistic data forwarding.

(2) Calculating Centrality with Local Information
   Centrality in graph theory and network analysis is a quantification of the relative importance of a vertex in the graph. It is a nature measure of the structural importance of a node in the network.
   In ZOOM, we use *betweenness* [70] to measure the centrality of vehicles, which refers to the extent to which a vehicle lies on the social paths linking other vehicles. Therefore, a vehicle with a high betweenness has a capability to facilitate interactions between the vehicles it links. With only local information, we adopt the algorithm [71] to calculate the betweenness in *ego networks*, which refers to a network consisting of a single vehicle (ego) together with the vehicles (alters) the ego is connected to and all the links among those vehicles. Although the betweenness in ego networks does not correspond perfectly to the global betweenness, the ranking of vehicles are identical in the network. Mathematically, we present the relationships between an ego vehicle v_i and its neighbors in the ego network by a $m \times m$ symmetric matrix A,

$$A_{ij} = \begin{cases} \theta_{i,j}, & \text{if} \quad \theta_{i,j} > 0 \\ 0, & \text{otherwise} \end{cases}. \qquad (4.13)$$

where m is the number of neighbors and $\theta_{i,j}$ is the regularity ratio between v_i and v_j. The ego betweenness of v_i can be calculated as the sum of the reciprocals of the entries of $A^2[1 - A]_{i,j}$ [71]. The ego betweenness of a vehicle is updated upon each contact with other vehicles. Specifically, when vehicle v_1 meets v_2, v_2 sends a list of neighbors in its ego network to v_1. Upon receiving the neighbor list, v_1 checks each neighbor in the list, $v_i, i \in [1, 2, \ldots, l]$, if v_i is also a neighbor of v_1, then elements $A_{2,i}$ and $A_{i,2}$ are set to $\theta_{2,i}$. If v_2 is a newly encountered vehicle, v_1 will first enlarge $A_{m \times m}$ to $A_{(m+1) \times (m+1)}$ by inserting a new row and a new column for v_2. Then it performs the ego betweenness calculation accordingly. Vehicle v_2 conducts the same operations as v_1 at the same time.

4.3.3.4 Opportunistic Forwarding Strategy

In ZOOM, when vehicle v_1 encounters v_2, v_2 will send a list of all its neighbors and a list of destinations of packets it is currently carrying to v_1. Vehicle v_1 then update the Markov chain and calculates its ego betweenness. For the destination v_d of a packet of v_2, let $\underline{b}_{v_1,v_d}$ denote the current state in the k-th order Markov chain between v_1 and v_d. The estimated delay of the next contact between v_1 and v_d, $\mathcal{E}_{delay}^{v_1,v_d}$ can be calculated as,

$$\mathcal{E}_{delay}^{v_1,v_d} = \sum_{a=0}^{\lfloor T/\lambda \rfloor} p_{b_{v_1,v_d};a} \cdot a. \qquad (4.14)$$

Vehicle v_1 will act as the next relay for this packet if one of the three following cases happens: 1) v_1 is the destination of this packet, i.e., $v_1 = v_d$; 2) v_1 has a shorter estimated delay of the next contact between v_1 and v_d than that between v_2 and v_d, i.e., $\mathcal{E}_{delay}^{v_1,v_d} < \mathcal{E}_{delay}^{v_2,v_d}$ and 3) both v_1 and v_2 have no prior about v_d and v_1 has a larger betweenness value than v_2. After transmitting the packet to v_1, v_2 removes this message from its buffer. Similarly, v_2 conducts the same operations accordingly.

4.4 Performance Evaluation

4.4.1 Methodology

In this section, we compare our opportunistic forwarding algorithm with several alternative schemes:

- **Epidemic**. In this scheme [46, 47], vehicles exchange every packet whenever they experience a contact. If vehicles have infinite buffer size, using epidemic routing will find the shortest path between the source and destination vehicles and therefore has the shortest end-to-end delay. On the other hand, since there is no control on data forwarding, it also generates a tremendously large volume of network traffic, overwhelming limited wireless bandwidth.
- **Markov**. This scheme [54] establishes a kth order Markov chain to predict the time when the next contact may occur between a pair of vehicles, utilizing the temporal correlations of consecutive ICTs. A greedy strategy is taken in making routing decisions where the neighboring vehicle with the least estimated meeting time with the destination will be chosen as the next data relay.
- **SimBet**. This scheme [55] assesses similarity between nodes in a social graph to detect nodes residing in the same community, and betweenness centrality to identify bridging nodes which could carry a packet from one community to another. Packets are routed to the most central nodes until a node with higher similarity with the destination is met. Then the packet is forwarded within the community until the destination is reached.
- **Bubble Rap**. This scheme [56] uses a similar approach as SimBet except that communities here are explicitly identified by a detection algorithm.

We consider four important metrics to evaluate the performance of ZOOM and the above schemes:

(1) *Delivery ratio*. It refers to the ratio of successfully delivered packets to the total number of packets at the end of an experiment.
(2) *End-to-end delay*. It refers to the delay for a packet to be received at its destination. We only calculate end-to-end delay for successfully delivered packets.
(3) *Network traffic per packet*. It refers to the average network cost per packet, calculated by dividing the total number of data forwarding hops by the total number of packets.
(4) *Packet utility*. It refers to the average benefit in reducing the delivery delay by each forwarding hop, calculated by dividing the total amount of time saved (i.e., the time period starting since a packet is delivered and ending when the experiment ends) for all packets to the total number of data forwarding hops.

In the following simulations, we evaluate the above metrics of ZOOM, using real trace data of Shanghai buses for demonstration. We randomly choose 1,000 buses, and use the contact records of 3 weeks from Feb. 19 to 28, 2007 for the initialization of all alternative schemes and use contact records of four and a half days from 8 am on Mar. 1 to 5, 2007 for data forwarding experiments (the reason that we set the experiment to start from 8 am in the morning is because most buses are not in service at night.). At the beginning of each experiment, we inject 100 packets using a Poisson packet generator with a mean interval of 10 s. For each packet, the source and destination are randomly chosen among all buses in the data set. Here we make a general assumption that two vehicles can always successfully conduct all data transmission when they have a contact. We run each experiment 50 times and get the average.

4.4.2 Performance Comparison

In this simulation scenario, we compare ZOOM with all other alternative forwarding algorithms. For the sake of fairness, we adjust the contact aggregation scale for the best delivery performance for SimBet and Bubble Rap. In this simulation setting, the optimal number of contacts for a pair of vehicles to have a link in the contact graph is twenty.

Figure 4.23 plots the average delivery ratio as a function of experiment time. It can be seen that ZOOM outperforms other algorithms except the epidemic routing. As epidemic routing can always find the shortest path by aggressively spreading a packet over the whole network, it also causes unacceptable network traffic. It can be seen that ZOOM is capable of obtaining great delivery ratio gain in a very short period of time. For instance, ZOOM can successfully deliver over 60 % packets within 24 h while the ratio for Markov, SimBet and Bubble Rap is 35, 37 and 24 %, respectively. In addition, it is very interesting to see that, for all schemes, the delivery ratio stabilizes and stops to increase when it is night, for example, during the first night from the 14th hour (i.e., 10 pm on Mar. 1) to the 22th hour (i.e., 6 am on Mar. 2), and the second night from the 38th hour (i.e., 10 pm on Mar. 2) to the 46th hour (i.e., 6 am on Mar. 3). The reason is that the data forwarding process would suspend during the night as most buses are not in service at night and would continue in daytime when buses are on duty.

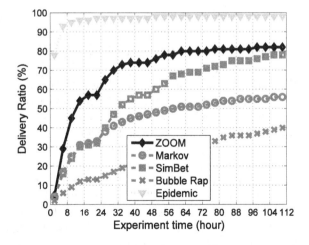

Fig. 4.23 The average delivery ratio versus the experiment time

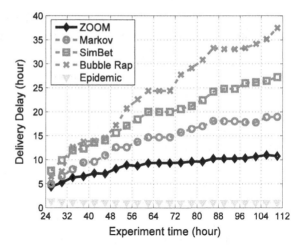

Fig. 4.24 The average delivery delay versus the experiment time

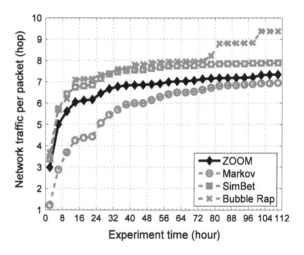

Fig. 4.25 The network traffic versus the experiment time

Figures 4.24 and 4.25 plots the average end-to-end delay and network traffic as a function of experiment time for all successfully delivered packets, respectively. Note that, for comparison fairness, we only take into account those packets that can be successfully delivered by all schemes. It is clear to see that, in general, algorithms utilizing contact-level mobility can achieve very small delivery delay comparing with social-level-mobility-based routing algorithms. Moreover, ZOOM can achieve the minimum end-to-end delay (excluding the epidemic routing). It can be seen that Markov generates the least network traffic and ZOOM introduces slightly more traffic than Markov. Combining both the end-to-end delivery delay and the network traffic, we argue that ZOOM can actively spend few more hops to achieve far more gain in end-to-end delivery delay. In contrast, schemes based on social-level mobility spend more hops but result in larger delays and, therefore, have less network-cost-efficiency.

Fig. 4.26 The packet utility versus the experiment time

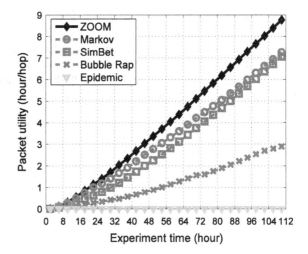

In general, routing algorithms try to trade off between delivery delay and network traffic cost. Short delivery delays usually imply large network traffic. To better measure how efficient a routing algorithm can be, we evaluate all schemes with the packet utility metric. Figure 4.26 plots the average packet utility as a function of experiment time. It is clear that ZOOM has the highest packet utility among all schemes. In summary, ZOOM is a very fast and cost-efficient opportunistic routing scheme under urban VANET settings.

4.5 Summary

In this chapter, we have first demonstrated that urban vehicles show strong temporal dependency in terms of how they meet each other. Although our study based on two specific types of public vehicles, namely taxies and buses, they are representative with respect to mobility characteristics in urban settings. Buses have dedicated routes and fix schedules which make their connection opportunities more predictable. On the other hand, taxies with much random mobility still have strong temporal correlation between every pairwise contact.

Then, we have presented an appealing contact-level opportunistic forwarding algorithm using higher order Markov chains, which can significantly improves the delivery ratio and reduce the end-to-end delay for data delivery. Furthermore, we have proposed an opportunistic forwarding algorithm, ZOOM, which captures both lower level mobility at pairwise contact scale and upper level mobility from the VANET perspective. ZOOM uses locally collected contacts to predict the future contact opportunities between vehicles. Moreover, the capability to predict contacts is then utilized to reflect the social relation ties between vehicles. We have demonstrated the efficacy of our algorithms through extensive trace-driven simulations.

We will extend our work in two directions in the future. First, it is often assumed in the literature that data transfers can be done instantaneously as soon as two vehicles have a chance to meet. It is definitely not the case in reality since wireless link quality is very dynamic. Thus, we will investigate the end-to-end delay with limited wireless link bandwidth since the delay is influenced not only by ICTs but also by retransmissions if the data transfer fails in a contact. Second, we will validate our algorithm by conducting field tests and collecting trace data of more types of vehicles.

Chapter 5
Vehicle Tracking

5.1 Introduction

Among the others, online real-time vehicle tracking is a fundamental service in the SG project, which refers to tracking the current position of a certain vehicle in real time. A wide spectrum of compelling applications can be implemented on top of this basic service. For example, authorized users will be able to track individual vehicles that they are concerned with, such as their own or friends' cars, public buses and taxies. In particular, there exist several critical types of vehicles in the city, which need to be located urgently, such as stolen cars, speeding cars, ambulances and police cars. Besides these application scenarios, it is also an indispensable building block underpinning many other high-level applications. For example, in the bus arrival prediction application, the tracking service is used to locate the nearest feasible bus.

However, real-time vehicle tracking in the metropolitan-scale system is very challenging because of several rigorous requirements. First, users (or high-level applications) often pose a real-time requirement on tracking a certain vehicle. That is, any query for the vehicle must be answered within a certain bounded time. Otherwise, the returned answer may become invalid or useless. For example, a query tries to locate the current location of a stolen car. If the query fails to be answered within a short time, the car could actually be far away from the returned location because it may be moving at a high speed. Second, the system should be scalable to support hundreds of thousands of vehicles. In addition, SG aims to serve millions of users every day. The huge number of simultaneous queries is a serious issue. In addition, as the Shanghai city is continuously expanding, the system is required to be highly extensible to such expansion. Third, the system should be robust to node failures. In such a large-scale distributed system consisting of thousands of local nodes, system maintenance is not a trivial issue.

To realize this service, a centralized scheme is straightforward, where location information of all vehicles is sent back to a centralized database and constantly maintained. A user, who wants to track a vehicle, can send a query to the central server. The server then processes the query and returns the location information of

H. Zhu and M. Li, *Studies on Urban Vehicular Ad-hoc Networks*,
SpringerBriefs in Computer Science, DOI: 10.1007/978-1-4614-8048-8_5,

that vehicle to the user. However, it is infeasible for the metropolitan-scale system due to the huge amount of vehicle data streams. For example, there are 22,413 crossroads in Shanghai. Even in 2001, the average number of vehicles running across a crossroad per minute in daytime was up to 33 [77]. This produces in total about 12 thousand events per second. Such a large volume of location updating data can easily overwhelm the centralized server. Therefore, it necessitates efficient designs of distributed solutions. As an alternative scheme, captured vehicle information can be stored locally at distributed nodes. As a result, there is little updating data as in the centralized scheme. Nevertheless, by this means, there is no hint about the enquired vehicle for a query. To track the vehicle, an intuitive scheme is to flood the query across the network which can always locate the desired vehicle. However, flooding search incurs a large amount of network traffic and hence is subject to poor scalability. To reduce query traffic, there is another search scheme based on random walks, which introduces modest network traffic. But, this scheme is limited by the problem of unbounded response latency of the query. As a result, there is no existing successful solution, to the best of our knowledge, to tracking vehicles in real time in a large-scale distributed system.

In this chapter, we propose a novel scheme Hierarchical Exponential Region Organization (HERO) [75, 76] which satisfies the unique requirements of real-time vehicle tracking in a metropolitan-scale distributed system. In SG, wireless access points (APs) and RFID readers will be deployed throughout Shanghai. Exploiting the pervasive deployment of these devices, location and status information of vehicles can be actively captured and logged in a large number of distributed local nodes. In essence, HERO connects local nodes into an overlay network matching the underlying road network. A hierarchical structure over the overlay network is constructed and dynamically maintained while the vehicle is moving along. Exploiting the inherent spatiotemporal locality of vehicle movements, this hierarchy enables the system to conservatively update location information of a moving vehicle only in nearby nodes. The distinctive features of HERO are twofold. First, it guarantees that any query, which can be injected anywhere in the city, can meet the real-time constraint associated with each vehicle, by bounding the maximum number of hops that the query is routed. Second, it significantly reduces the communication overhead of both location updating and query routing, and therefore is truly scalable to support hundreds of thousands of vehicles and millions of system users. Moreover, HERO is a fully distributed light-weight protocol extensible to the increasing scale of the system. In addition, it is robust to node failures and able to tolerate inaccurate location readings.

The remainder of this chapter is structured as follows. Section 5.2 compares HERO with related work. In Sect. 5.3, we introduce the infrastructure that will be deployed in the SG project. Section 5.4 elaborates the design of HERO and presents theoretical analysis for the optimal configuration of the protocol parameters. Several design issues that may be encountered in practice are discussed in Sect. 5.5. Section 5.6 describes our prototype implementation of the vehicle tracking system realizing the HERO protocol. In Sect. 5.7, we introduce the trace-driven methodology that we use to evaluate the performance of HERO and present simulation results. Finally, we present concluding remarks and outline the directions for future work in Sect. 5.8.

5.2 Related Tracking Approaches

Using GPS to localize and track vehicles is a straightforward solution. Several crucial reasons prohibit this solution for vehicle tracking in cities like Shanghai. First, with crowded high buildings squeezed along most of the narrow streets in the city, it is very difficult for the GPS system to work accurately without any other assistant devices. It is often the case that the reported GPS position of a vehicle can be more than 100 m deviated from its actual location. To make things worse, a large number of major roads are covered by viaducts which prevent satellites from seeing the vehicles running under them. Second, the intervals of location information reports can be notably long. Due to the GPRS communication cost for transmitting the GPS location information back to the data centre, drivers prefer to choose relatively large intervals. The typical value would be from one minute to three minutes. Third, the expense of a GPS receiver as well as data communication cost is quite high, which limits the wide deployment of this technology.

The Globe system [78] has constructed a static world-wide search tree for mapping object identifiers to the locations of moving objects. It is not flexible to expand or adjust the structure and may have the bottleneck problem near the root of the directory tree structure. In [79], the authors have introduced a distributed approach for load balance but they have not taken the number of system users into consideration. In contrast, HERO needs no dedicated directory servers and achieves good scalability and flexibility.

In database community, indexing techniques have been proposed for tracking moving objects but they are based on the assumption of the existence of centralized databases [80–83]. Despite the large number of existing methods, there is no applicable one for update-intensive applications, where it is infeasible to continuously update the index and process queries at the same time [84]. HERO does not need any centralized database and all routing information is distributed to every node in the system.

In structured peer-to-peer (P2P) networks, various DHT schemes have been proposed to map objects to peers in a decentralized way, thus enabling the system to satisfy queries efficiently [85–88]. However, DHTs may cause large computation and traffic overhead for a large number of rapid updates of moving objects. In unstructured P2P networks, the most typical query methods are based on flooding [89]. Using flooding is not scalable. Several randomized approaches, such as random walks [47, 48] and randomized gossip-based methods [90, 91] have been introduced to distribute and locate objects. Random walks are resilient to node failures but need sufficiently long walks before finding the results in a stable network. Random gossip-based methods can retrieve global information with high probability after approximately logarithmic rounds but introduce large traffic. Furthermore, none of these schemes provides real-time guarantees for queries. HERO introduces minimal updating cost to guarantee the real-time constraints desired by the applications.

5.3 System Description

As RFID technology continuously evolves, it has been widely used in tracking various mobile objects, such as vehicles [92, 93]. The US government also enacts the TREAD Act [94] which demands RFID tags to be planted in every new tire before September 2007. The SG project exploits the promising RFID and local-area wireless communication technologies. The infrastructure of SG, which is still underway, is illustrated in Fig. 5.1. RFID readers and wireless APs will be deployed throughout the urban area of Shanghai, typically installed at crossroads. A local node is responsible for collecting data from several close RFID readers and wireless APs within its own domain, and accepts queries from nearby users or applications. A local node is basically a server which connects to a dedicated underlying network for communication.

In SG, the vehicles' information is gathered both actively and passively. In the initial prototype of SG, a vehicle is captured passively using active RFID technology. An active RFID tag emits its ID at a fixed interval and has an effective

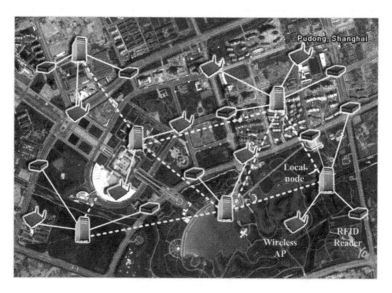

Fig. 5.1 The infrastructure of ShanghaiGrid; a small part of the Pudong District of Shanghai is shown

communication range of about 2–80 m. The battery can sustain the operation of an active RFID tag for about 6 years [95]. A moving vehicle attached with an active RFID tag can be captured if the emitted signal reaches some reader. Besides active RFIDs, a vehicle can actively communicate with wireless APs as it passes by them. A Cisco Aironet 1240AG access point working under IEEE 802.11 g has an effective outdoor communication range of about 280 m at the transmission rate of 2 Mbps [96]. The vehicle can actively push important vehicle status information, such as vacancy status, to local nodes.

Precisely speaking, we aim at providing real-time guarantee of tracking a vehicle by bounding the maximum number of hops that a query could traverse in the system. Since the provision of such a real-time service depends on the underlying network for communication, a dedicated network such as an ATM can be used which provides a reliable and predictable data transmission between any two endpoints. With the bounded maximum number of transmission, such a system for the purpose of tracking vehicles can guarantee rigid real-time requirements. In the initial prototype of SG, we connect local nodes to the wide-area ATM network provided by Shanghai Telecom [97] through a dedicated connection or a cheap ADSL connection.

5.4 Design of HERO

In this section, we first give an overview of the HERO protocol, introducing its basic rational. Next, we delve into the conservative location updating based on the assistance of a dynamically maintained hierarchy. Finally, we discuss the optimal configuration of the protocol parameters of HERO.

5.4.1 Overview

To meet the rigid requirements in vehicle tracking in real time, we need to solve two critical issues. First, the system should limit the maximum query response time to guarantee the real-time constrains from applications. Second, the system should minimize network traffic to support a large number of vehicles and queries as well as the continuous extension of the network.

However, there is an intrinsic tradeoff between network traffic and query response time in vehicle tracking. As mentioned earlier, by aggressively updating location information of a vehicle to all the other nodes, the system provides minimal query response time whereas introducing high updating network traffic overhead. In contrast, the system suffers from long query response time if the system does not perform any location updating. In general, more rigid real-time requirement on tracking a vehicle implies higher network traffic overhead.

HERO elegantly manages to solve the two critical issues in an integrated way. The core idea of HERO is to dynamically update location information of a moving vehicle to all the nodes in the system *in a controlled way*. Generally, the nodes closer to the vehicle are updated more frequently than those further from it and, therefore, have more accurate information about the current location of the vehicle. By this means, HERO effectively exploits the inherent spatiotemporal locality of vehicle movements in an urban setting, and consequently reduces location updating cost. Upon receiving a query, the node unlikely has the exact information. However, it knows some other node which has more accurate information about the vehicle. Thus, it forwards the query to that node. Following an elaborately organized routing path, the query can eventually reach the destination node, which keeps the most updated information of the vehicle. The typical latency between two nodes can be easily measured. Thus, by bounding the maximum number of hops that the query is routed, HERO can also meet the real-time constraint for the vehicle.

The key to the design of HERO is how to realize the controlled location updating while bounding the maximum number of hops a query is routed. To accomplish this, HERO integrates four effective components:

Overlay construction: To exploit the locality of vehicles' movements, HERO organizes local nodes into an overlay network that matches the real underlying road network in Shanghai (as depicted in Fig. 5.1, dashed lines present the overlay connections of local nodes). There is a connection between two geographically adjacent local nodes in the overlay network if there is a road between the two corresponding regions. This overlay is easy to build and maintain, with each node having to know its neighbors. Additional overlay connection may also be added for two nodes that are geographically close to each other even if they are not connected by a real road. Such connections enhance the reliability of the overlay network when a local node has only one road connecting itself to other local nodes.

Hierarchy organization: For every vehicle, HERO divides local nodes into different regions which constitute the hierarchy on the overlay network. The regions are organized in the following way, as illustrated in Fig. 5.2. The first region (R_1) has the smallest size and covers the vehicle. For the example in Fig. 5.2, R_1 covers node e, which is closest to the vehicle and has the latest

Fig. 5.2 Illustration of hierarchical regions and query routing

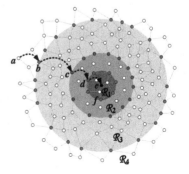

information about it. The second region (R_2) has a larger size and covers R_1. More generally, a region (R_i) has a larger size than the immediate inner region (R_{i-1}) and covers it.

Restricted location updating: When the vehicle is moving within R_1, the location updating involves only the small set of nodes in R_1. When the vehicle is moving out of R_1, the location updating is extended to more regions. In this case, part of the hierarchy needs to be re-organized. This reorganization aims to restrict location updating in R_1 as much as possible, thereby minimizing network traffic cost for location updating.

Query Routing: With the hierarchy and restricted location updating, a region always has more up-to-date location information of the vehicle than its outer regions. In HERO, each node has a pointer pointing to a boundary node of its immediate inner region. A query can be injected from any node in the system. For example in Fig. 5.2, node a receives a query. Node a will forward the query to b. The query will further be forwarded by nodes c and d, and eventually arrives at e. Node e will return the location information directly back to node a. To restrict the maximum number of hops that the query is routed, we limit the number of regions that the hierarchy for the vehicle contains.

In the following subsections, we first describe the process of hierarchy initialization when a new vehicle is joining the system. Next, we describe the detailed mechanism for restricted location updating while the vehicle is moving based on the established hierarchy. Finally, the optimal configuration of design parameters is discussed.

5.4.2 Hierarchy Initialization

The first node that captures a new vehicle triggers an initialization procedure to establish the hierarchy for the vehicle. As the vehicle may move towards any direction, a region is initially designed as a disk in the overlay network. Note that, the deployment of local nodes is not necessary to be uniform in the city. They can be more densely deployed where more refined tracking accuracy is required. We will discuss more on this in Sect. 5.5. In the rest part of this chapter, without explicit specification, distance is measured in terms of hops in the overlay network. Each region R_i has a radius r_i (in hops). A node, which has a distance d from the first node, belongs to region R_k if this region is the smallest one that covers the node. The radius r_k of R_k is,

$$r_k = \min_{i=1}^{h}\{r_i, r_i \geq d\}, \tag{5.1}$$

where h is the maximum number of regions in the system. If d equals to certain r_i, $1 \leq i \leq h$, the node is on the boundary of R_i. Moreover, for query routing, every

Table 5.1 Data structures used in the algorithm	Local node	Init. packet	Update packet
	Boundary	Router	Router
	Next-insider	Journey	Journey
			Scale

node maintains a pointer that points to a node which is on the boundary of the immediate inner region. Let next-insider denote this pointer.

To establish next-insider pointers in the nodes, the first node initiates an initialization packet which contains a router field for setting up these pointers and a journey field for maintaining the distance that the packet has traversed. The first node initializes router and journey to its own IP address and one, respectively. Then the first node floods the initialization packet throughout the network. Upon receiving the packet, a node first sets its next-insider to router contained in the packet. Then it checks journey in the packet. If journey equals to the radius of certain region r_i, the node marks itself as a boundary node of region R_i. It also modifies router in the packet to its own IP. Otherwise, it leaves that field unchanged. Next, it increases journey in the packet by one and re-broadcasts the packet to its neighbors. In addition, duplicated initialization packets with larger journey are silently dropped. After the initialization procedure terminates, the regions are centered at the first node and the hierarchy is established (as illustrated in Fig. 5.2). Note that the structure of the hierarchy is distributed in all local nodes (the data structure for a node is shown in Table 5.1). Therefore, the storage overhead for tracking the vehicle at a local node is very small.

5.4.3 Restricted Location Updating

When a vehicle is moving in the city, its information is captured by the local nodes that it passes by. When a node captures the vehicle (we call this node *chaser*), it performs location updating, and maintains the hierarchy for the vehicle if necessary. There are three cases. For presentation clarity, we define a node as a *boundary node* of R_i if it is a most outer node within R_i. The nodes in R_i except boundary nodes are *interior nodes* of R_i.

Case 1: the chaser is an interior node within R_1. In this case, the hierarchy for the vehicle remains unchanged. The chaser floods the location information of the vehicle to all the other nodes in R_1.

Case 2: the chaser is a boundary node of R_1. In this case, it is possible that the vehicle will move out of R_1 shortly. For example in Fig. 5.3, node a is the current chaser which is a boundary node of R_1' (the dashed circle). When the vehicle moves along the depicted direction, R_1' will not cover the vehicle any more. Two consequences follow. First, a future query cannot be routed to the chaser properly

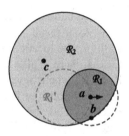

Fig. 5.3 Reconstruction of R_1, node a is the chaser and is a boundary node of the first region (R_1')

because the information on the boundary nodes of R_1' is out-of-date. Second, to enable the proper routing of a future query, the chaser has to flood the location information of the vehicle to R_2 every time, which will incur larger network traffic overhead. Therefore, HERO needs to re-organize R_1.

To this end, the chaser initiates an *update packet* in which its **router** and **journey** is initialized to its own IP address and one, respectively, as in an initialization packet. The update packet includes an additional **scale** filed that is used to indicate the area that the update packet should be propagated to. In this case, the chaser floods the packet within R_2 by letting the boundary nodes of R_2 stop the flooding. On the one hand, the new R_1 is rebuilt from the current chaser within R_2. At the same time, location information is also updated in the new R_1. On the other hand, it updates nodes in R_2 about the current position of the new R_1.

There is a special situation during the reconstruction of R_1, where the new R_1 is truncated by the boundary of R_2. This happens when the chaser is close to the boundary of R_2 (e.g., node a in Fig. 5.3). In this situation, a boundary node of R_2 receives an update packet whose journey is less than or equals to r_1 (e.g., node b in Fig. 5.3). As a result, this node sets itself as a boundary node of both R_1 and R_2. We call such a node a common boundary node of R_1 and R_2. In this case, R_1 is no longer a disk because it is restricted in R_2. But, this does not affect the operation of our protocol.

Case 3: the chaser is a common boundary node of several regions R_1, R_2, ..., R_j ($j > 1$). This is actually a more general situation of Case 2. This situation results from constant reconstructions of regions as the vehicle is moving. In this case, it is possible for the vehicle to move out of all the regions from R_1 to R_j. The system needs to re-organize regions from R_1 to R_j. For example in Fig. 5.4, the situation occurs if node b is the current chaser, where b is also a common boundary node of R_1' and R_2' (the dashed circles).

Fig. 5.4 Reconstruction of R_1 and R_2, node b is the chaser and also is a common boundary node of the first and second region (R_1' and R_2')

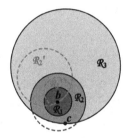

To re-build regions from R_1 to R_j, the chaser floods an update packet within R_{j+1}. As a result, all regions from R_1 to R_j are re-constructed within R_{j+1}. In addition, the location information of the vehicle within R_{j+1} is updated. Similar to Case 2, there is also a special situation during the reconstruction of regions from R_1 to R_j, where several regions, say from R_i to R_j, might be truncated by some boundary nodes of R_{j+1}. Such a boundary node of R_{j+1} sets itself as the common boundary node of regions R_i, R_{i+1},..., R_{j+1}, ($1 \leq i \leq j$). For example in Fig. 5.4, node c is a resulting common boundary node of R_2 and R_3.

Note that the hierarchy needs to be established only once at the time when the vehicle is first introduced in the system. Afterwards, it is dynamically maintained in a fully decentralized manner. Therefore, the storage overhead for tracking the vehicle at each local node is small. HERO automatically reorganizes the hierarchy to control the flooding for location updating to happen mostly in the first few smallest regions. Using flooding for the controlled location updating and hierarchy maintenance is robust and effective when the flooding scale is small [98]. In addition, duplicated useless packets during the flooding are silently dropped which also mitigates the network traffic for location updating. The efficacy of HERO design can be examined more intensively by our prototype system implementation and extensive simulations.

5.4.4 Protocol Analysis and Parameter Optimization

By far, a key question remaining unestablished is the configuration of the radii r_i ($1 \leq i \leq h$) in (5.1). To conveniently control the maximum number of regions in the hierarchy and to restrain the location updating in small regions close to the vehicle, HERO organizes the hierarchical regions with exponentially increasing sizes.

More specifically, we introduce two protocol parameters: first radius r and amplification factor k. The radius of the first region is r (i.e., $r_1 = r$), and the radius of R_i is $k^{i-1}r$ (if k is an integer). Figure 5.2 shows an example with r and k both equal to 2. More generally, k can take any real number greater than one. Since the radius of a region must be an integer in hops, we take the ceiling of $k^{i-1}r$ as the radius of R_i and further make sure that a region is larger than its immediate inner region. Then the radius of R_i is defined as,

$$r_1 = r;$$
$$r_i = \begin{cases} \lceil r \cdot k^{i-1} \rceil, & \text{if} \quad r_{i-1} < \lceil r \cdot k^{i-1} \rceil; \\ r_{i-1} + 2, & \text{otherwise.} \end{cases} \qquad (5.2)$$

We are interested in the maximum number of hops that a query is routed, and we have the following theorem.

Theorem 1 *Given a network with the network diameter (i.e., the maximum hop distance between any pair of nodes) D hops, it takes at most $\lceil \log_k(D/r) \rceil$ hops for a query to be answered.*

Proof The worst case of a query, where it traverses the maximum number of hops, occurs when the hierarchy is constructed from one end of the network diameter and the query is injected at the other end of the diameter. In this case, according to the definition of the exponential hierarchy, the maximum number of regions contained in the network is $\lceil \log_k(D/r) \rceil$. Since nodes in R_1 always have the latest location information, a query only needs to be routed to a boundary node of R_1. Thereby, a query takes at most $\lceil \log_k(D/r) \rceil - 1$ hops to reach that boundary node. It takes the boundary node one more forwarding hop to finally return the result back to the node that initiates the query. This concludes the proof. ∎

We study the location updating overhead caused by the movements of a vehicle. Since the patterns of the vehicles' movements could be very different, we analyze the updating overhead in the worst case where a vehicle moves straight. In this case, the movement continuously breaks the maximum number of regions, and therefore arouses the most significant updating overhead. We have the following theorem.

Theorem 2 *Suppose that the topology of a network is a disk, the maximum network traffic overhead of location updating for a vehicle moving a distance of D is $\eta(D) = c(kD^2 + 2r(r-k-1)D - 6r^2)$, where D is the network diameter and c is a constant coefficient.*

Proof Figure 5.5 depicts the worst case of location updating among all possible movements with a distance of D, where all constructed regions in the network need to be reconstructed during the movement from node a to node b. For analysis simplicity, we assume that k is an integer. With uniform deployment of local nodes, the network traffic for flooding in R_i (denoted as S_i) can be approximately evaluated by the area of R_i. Let ζ_i denote the updating overhead incurred as a vehicle moves from the boundary of R_{i-1} to the node immediately next to the boundary of R_i, ($i \geq 2$). For example in Figure 5.6, the updating overhead introduced when the vehicle moves from node a to node b is denoted as ζ_1, and that from node c to node d is denoted as ζ_2. We have,

Fig. 5.5 Worst case of location updating, when the vehicle traverses the whole network from node a to b

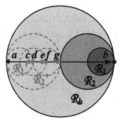

Fig. 5.6 Example of
continuous reconstruction of
R_1 during the movement from
node a to d

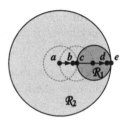

$$\begin{cases} \zeta_1 = (r-1)S_1 = c_0\pi(r-1)r^2 \\[2mm] \zeta_i = (k-1) \cdot \left(S_i + \sum_{j=1}^{i-1} \zeta_j \right) \end{cases} , \tag{5.3}$$

where c_0 is a constant coefficient. Let ω_m denote the updating overhead incurred as the vehicle traverses the diameter of R_m from node a as shown in Fig. 5.5.

To formulate ω_m, we need to go through the whole process of the restricted location updating when the vehicle moves from one end of R_m to the other. Obviously, we can recursively express ω_m in terms of ω_{m-1} and ζ_{i-1},

$$\omega_m = \omega_{m-1} + 2(k-1)(S_m + \zeta_{m-1}). \tag{5.4}$$

For example in Fig. 5.5 where k is 2, ω_2 (i.e., the updating overhead incurred when the vehicle moves from node a to node g) consists of ω_1 (from node a to node c), S_2 (from node c to node d), ζ_1 (from node d to node e), S_2 (from node e to node f) and ζ_1 (from node f to node g).

Thus, with Eqs. (5.3) and (5.4), ω_m can be formulated as follows,

$$\begin{aligned} \omega_m &= (2r-1) \cdot S_1 + 2(k-1) \left(\sum_{i=2}^{m} S_i + \sum_{i=1}^{m-1} \sum_{j=1}^{i} \zeta_j \right) \\ &= c_0\pi \left(2k \cdot r_m^2 + 2r(r-k-1)r_m - 3r^2 \right) \end{aligned} \tag{5.5}$$

Denote $\eta(D)$ as the total updating traffic caused while the vehicle traverses the network, and then $\eta(D) = \omega_h$. Let $c = c_0\pi/2$. This concludes the proof. ∎

We aim to meet the real-time constraint of a vehicle and meanwhile minimize network traffic overhead. The typical latency between a pair of local nodes connected using ATM connections can be measured. Let t_d denote the maximum delay of a query between two adjacent nodes, and t_0 denote the application real-time constraint. We try to minimize the average updating overhead per hop, $\eta(D)/D$, under the constraint $\lceil \log_k(D/r) \rceil \le t_0/t_d$. The average is a function of r and k, and let $g(r,k)$ denote it. Then, we have,

$$\begin{aligned} g(r,k) &= \eta(D)/D \\ &= c(k \cdot D + 2r(r-k-1) - 6r^2/D) \end{aligned} \tag{5.6}$$

Fig. 5.7 First radius r as a function of D

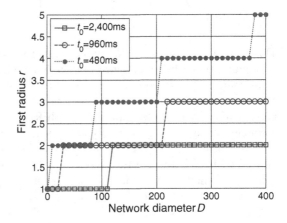

To minimize the traffic overhead, let $\log_k(D/r) + 1 = t_0/t_d$, and $g(r, k)$ can be reduced to,

$$g(r) = c\left[D^{\frac{t_d}{t_0-t_d}} \cdot r^{-\frac{t_d}{t_0-t_d}}(D - 2r) + 2\left(1 - \frac{3}{D}\right)r^2 - 2r\right]. \qquad (5.7)$$

Further, let the differentiation of $g(r)$ equal to zero, $dg(r)/dr = 0$. Since it is difficult to derive the exact r and k that produce the smallest network traffic overhead, we develop numerical procedures to compute the approximately optimal value of r and k. Figures 5.7 and 5.8 show the optimal values of r and k using numerical computation, respectively, where t_d is set to 48 ms in the example. It can be seen that the first radius of R_1, where HERO tries to restrain the locating updating, increases very slowly with the network scale.

Fig. 5.8 Amplification factor k as a function of D

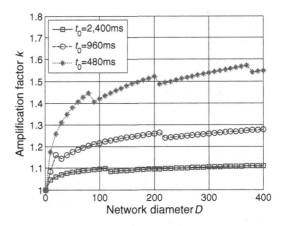

5.5 Design Issues

This section discusses some design issues that HERO may encounter in practice.

Scalability. HERO is designed to track hundreds of thousands of vehicles in a metropolitan-scale system with a large number of users. Therefore, the system scalability concern in terms of the number of vehicles, the number of users and the number of local nodes is critical. With HERO, the system needs to maintain a hierarchy for each vehicle. If every movement of a vehicle will introduce a lot of location updating traffic into the system, the cost can be prohibitively expensive. However, this is where HERO comes to help. HERO leverages the inherent locality of vehicle movements and only updates a small number of nodes nearby the vehicle. Therefore, the location updating cost should be small. We can also notice that the query cost is modestly low, which is a logarithmic scale to the size of the network. Moreover, each node in the system only needs to maintain the information of several neighboring nodes. It is a light-weighted protocol to join and leave the system. We will further investigate the scalability of HERO by extensive trace-driven simulations in Sect. 5.7.

Resilience to unreliable data. It is possible that occasionally a vehicle is not captured by an RFID reader (e.g., when the vehicle is moving too fast). In addition, a local node may also fail from time to time. It is critical to the operations of HERO if a boundary node misses a vehicle passing by. This inaccuracy can be easily detected in the system. At any time, a node in region R_i ($i \geq 2$) should have received an update packet from a boundary node of R_{i-1} before the node itself captures the vehicle. Otherwise, it is aware that the vehicle has escaped from R_{i-1} and the corresponding updating process fails. To solve the problem, we let the node which discovers this inaccuracy take the responsibility over as if it were a boundary node of R_{i-1} and triggers updating for the reorganization of regions from R_1 to R_{i-1}. Unless the node itself happens to be a boundary node of R_i, it performs updating for the reorganization of regions from R_1 to R_i instead.

Tracking accuracy. As a vehicle keeps moving, it may run out of the reading range of an RFID reader while still has not entered the territories of others. This causes the system have inaccurate vision about the current position of the vehicle before the vehicle re-enters into the system. It also defines the resolution of tracking accuracy of the system to be the uncovered distance between two adjacent RFID readers. In more practical environments, this inaccuracy can be enlarged when RFID readers fail to capture the vehicle as the vehicle passes. To refine the tracking resolution, more RFID readers can be deployed in the system. In order to reduce the cost, readers can be deployed more densely at those places where more accurate location information of individual vehicles is required and less densely at other places.

Node join and maintenance. In HERO, a single node failure can be discovered in a short time. A local node can periodically check with its neighbors while performing HERO protocol. An unavailable node is then reported to the system administrator. To join the system for tracking vehicles, a new node (or a recovered node) only needs to contact its adjacent nodes. Then, for each vehicle, the node configures its status the same as that of the neighbor which resides in the smallest region among all neighbors in the hierarchy. Thereby, it knows its position in the hierarchy for each vehicle and can perform location updating and query processing properly.

Data replication. The tracking data of vehicles can be of great importance for many applications. It is an important issue for the system to protect these data from node failures and disasters, such as fires or earthquakes. HERO actually has the implicit advantage of protecting important tracking data. Recall that tracking data are replicated in the first region. It implies that the system is still able to track the vehicle even when the chaser node becomes unavailable. If some vehicles are particularly important and need additional protection of tracking data, we can make r relatively large associated with the vehicle. By this means, more data can be replicated in the first region organized for the vehicle.

5.6 Prototype Implementation

To validate the HERO design and prove its practical implementation, we have built a prototype system in the campus to track experimental vehicles. This prototype system contains 45 local nodes distributed in our campus. As shown in

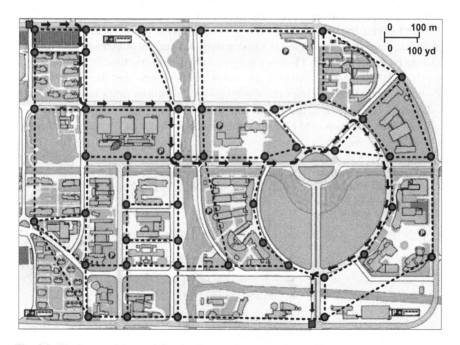

Fig. 5.9 The layout of the prototype implementation consisting of 45 nodes denoted by red spots

Fig. 5.9, local nodes (denoted by red spots) are deployed at crossroads of main roads. Every local node has an IEEE 802.11g wireless network interface connecting the local node to the campus Internet. Furthermore, the overlay network formed by the local nodes is illustrated by the dashed lines in Fig. 5.9. An overlay connection is established between two nodes if there is a road immediately connects them.

In the prototype system, we employ an active RFID system using "Tag Talk First" technology. Figure 5.10 shows a typical local node, which is associated with a SP-D300 RFID reader [10] as well as an IEEE 802.11g wireless AP. The inset of Figure 5.10 shows an active RFID tag (in highlighted area) attached to a vehicle. The reader's operating frequency is 2.4 GHz. It connects to the local node via a RS-485 interface and has a data transfer rate of 1 Mbps. The reader has a configurable operation range from 2 to 80 meters. Each reader can simultaneously detect up to 200 tags in 800 ms. Each tag has a unique 64-bit ID. Its battery has a life of 6 to 8 years. Tags send their unique ID signal in random with an average of 300 ms and can be detected at a high speed up to 125 miles per hour. Besides the RFID system, wireless communication technology is also investigated in our prototype implementation. The HERO protocol runs on Red Hat Fedora 5 and uses POSIX.1 socket API to communicate with each other. UDP packets are adopted for location updating and query routing. The size of all packets is 40 bytes, which includes 20 bytes of the IP packet header, 8 bytes of the UPD packet header and 12 bytes of data.

With this prototype implementation, we conduct a variety of experiments. Since we use the campus Internet as the underlying network, real-time guarantee seems to be non-trivial because the jitter (end-to-end round-trip time) can vary largely. To demonstrate this, we randomly choose two local nodes to measure the round-trip time by ping. Figure 5.11 shows the measured round-trip time from June 27 to June 30 in 2008. It can be seen that the round-trip time increases sharply from 7 to 11 pm at night. Moreover, the peak value can be almost four times larger than that at daytime. Nevertheless, the round-trip time is much more stable during the daytime. Thus, we choose to perform experiments with our prototype system from

Fig. 5.10 A local node with a RFID reader and a wireless AP; the highlight area in the inset shows an active RFID tag attached to a vehicle

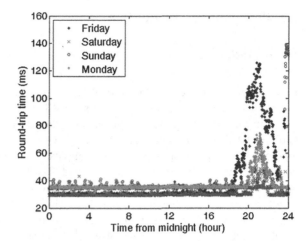

Fig. 5.11 The round-trip time pinged from two nodes randomly chosen from 45 nodes. The measurement is taken from June 27 (Friday) to June 30 in 2008 (Monday)

10 to 12 am on June 30 in 2008. We set the real-time constraint to 100 ms and take the maximum transmission delay between two online nodes which is 18.05 ms. Therefore, the resulting r and k are therefore configured to 3 and 1.278, respectively. We let a van carrying an active RFID tag travel at 30 miles per hour along the route as depicted by the dark arrows in Fig. 5.9. As the van enters an RFID reader's field and is captured by the reader, the associated node performs location updating accordingly. During the journey which lasts about 4 min, we let each node randomly generate one hundred of queries.

Among all the 4,500 queries, the maximum query latency is 90.45 ms, which is strictly shorter than the required real-time constraint. We also notice that the average query latency is about 47.93 ms. The network traffic for location updating among all nodes adds up to 13.2 KB. In contrast, the network traffic for location updating using broadcast on a spanning tree is about 28.8 KB. Since the maximum routing hops of a query is bounded (i.e., 5 hops in this experiment), the network traffic for query routing linearly increases with the number of queries in the system.

The lesson from our prototype implementation is that, with appropriate configuration of the protocol parameters, the query latency can be guaranteed to satisfy the real-time constraint requirement in terms of the number of hops that a message has to traverse. In addition, the overall network traffic overhead, introduced by location updating and query routing, can well accommodate a large number of queries. To further investigate the performance of HERO in a large-scale setting, we conduct trace-driven simulations, which are detailed in the following section.

5.7 Performance Evaluation

5.7.1 Methodology

In the simulations, the HERO protocol is implemented using ns2 [99]. Since we connect local nodes to the metropolitan-scale ATM network provided by Shanghai Telecom through cheap ADSL connections, the transmission delay between any two local nodes is reliable. Therefore, we can construct the overlay network topology by simply mapping the real complex road network of Shanghai where local nodes are deployed on every crossroad. The typical link transmission delay between two neighbor nodes in the overlay network is 48 ms, measured by ping between two desktop PCs with 1 MB bandwidth ADSL connections. One of the overlay topologies employed in our simulations is depicted in Fig. 5.12. The topology containing 1,000 nodes (denoted by small hollow dots) covers the geographical downtown area of Shanghai. The dark line shows the network diameter in the topology which is 55 hops. Upon the ATM network, we use UDP protocol for communicate with the packet size being 40 bytes.

To investigate the impact of the vehicle moving patterns to the HERO design, we use real GPS trace data of taxies which were obtained with GPS technology from August 2006 to October 2006. Taxies can move more randomly and extensively in the whole city and, therefore, have more sense to be considered. A typical trace of a taxi in the downtown area of Shanghai through daytime (on Aug. 13, 2006) is shown by solid dots in Fig. 5.12. It can be seen that, when the taxi is vacant, it cruises around within an area most the time seeking for passengers, as

Fig. 5.12 The topology of the downtown area of Shanghai with 1,000 nodes deployed at crossroads of this area

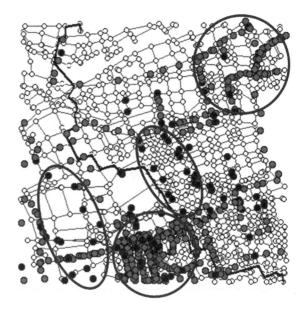

shown by these solid dots in the circle areas in Fig. 5.12. This benefits our HERO design best because most of the location updating can be perfectly restricted within small regions. It can also be seen that, when the taxi has a delivery, it runs very fast along the straightest path for its destination, as illustrated by those solid dots in the ellipse areas in Fig. 5.12. HERO leverages restricted location updating strategy to reduce network traffic while still keeping the whole system up-to-date.

We compare HERO with several alternative schemes:

- **ST-Updating**. In this scheme, whenever a node captures a vehicle, it updates this information to all other nodes. To reduce the network traffic overhead of this update, the system maintains a global spanning tree. Therefore, only $N-1$ update packets are introduced across the whole network for each update when there are N nodes in the network. The strength of this scheme is that each node can answer any query locally, providing minimal query response time.
- **ST-Flooding**. This scheme does not perform vehicle information update in the network and hence no overhead is introduced for location updating. To search for a vehicle, a query is flooded throughout the network. A global spanning tree is used to broadcast the query to reduce the network traffic overhead.
- **Ex-Flooding**. This scheme does not perform vehicle information update either. Without relying on a global spanning tree, it employs expanding flooding. The query is flooded in the overlay network. At the beginning, the TTL of the query is small. If this try is not successful, the query will be flooded again with an increased TTL (plus 4 hops). This process is repeated until the vehicle is found.
- **Random Walks**. Similar to ST-Flooding and Ex-Flooding, this scheme does not perform information update. To search for a vehicle, the query is carried out by five simultaneous random walkers. A walker checks with the querying node every 50 steps and terminates either if the querying node has already retrieved the result or if the maximum number of steps 2,000 is reached.
- **Chord**. In this scheme, each local node joins an overlay network of a logical ring [86]. With a series of indexing pointers maintained in each local node, each local node can update and retrieve the location information of a vehicle within $\log_2 M$ on average, where M is the size of the logical ring. In our implementation, M is set to 2^{32} which is moderate to support a large number of nodes and vehicles in the system.

We propose two important metrics to evaluate the performance of HERO and the above schemes:

(1) *Maximum query latency (MQL)*. It refers to the maximum query response time of a successful query. The intention of this metric is to check whether a scheme can guarantee certain real-time requirements.
(2) *Network traffic per query (MNT)*. It can be seen that if there were no query then no location updating would need to be carried out at all. Therefore, to answer a query, the system cost should involve two parts of network traffic, i.e., for location updating and for routing query packets. We investigate the

total communication cost per query, caused by any location updating as well as query processing.

5.7.2 Effects of Protocol Parameters

We first examine the effects of protocol parameters on the system performance and validate the theoretical analysis. We employ one hour extensive trace data of 100 taxies, randomly generate 10^5 queries for different vehicles during this hour and demand any query to be answered within 500 ms. We vary r from 1 to 30 hops with an increment of 1 hop, and vary k from 1.2 to 3 with an increment of 0.05. For each pair of r and k, we repeat the experiment 10 times and present the average.

Figures 5.13 and 5.14 plot M_{QL} among all the generated queries and M_{NT} under different configurations of r and k, respectively. It shows that M_{QL} drops dramatically with increasing r and k. It can be seen that M_{NT} increases with both increasing r and increasing k. This is reasonable because either a greater r or a greater k leads to a more aggressive updating strategy. At the extreme, if r equals to D or r equals to one and k equals to D, HERO floods every location updating throughout the whole network. In this experiment setting, according to the numerical computations in Sect. 5.3, r and k should take 2 and 1.393, respectively. The arrows in Figs. 5.12 and 5.13 show the corresponding positions. It is clear to see that, with this configuration of r and k, HERO can actually guarantee any query to satisfy the real-time requirement and meanwhile minimizing the network traffic overhead per query.

Fig. 5.13 Maximum query latency versus different protocol parameters

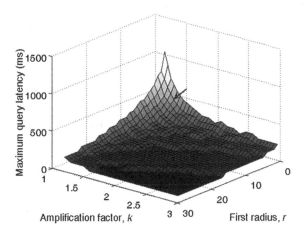

Fig. 5.14 Network traffic per
query versus different
protocol parameters

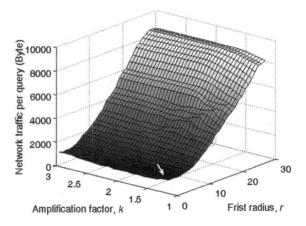

5.7.3 Impact of Query Quantity

In this experiment, we investigate the impact of the query quantity on the system performance. We take the same setting as the previous experiment. The protocol parameter r and k are set to 2 and 1.393, respectively. We vary the total number of queries from 10^3 to 10^5 with an increment of 400.

Among all queries, M_{QL} of HERO is 480 ms which is strictly shorter than the real-time constraint. In ST-Updating, M_{QL} is about 14 ms which is for local database operations. The other schemes cannot guarantee to satisfy the real-time requirement. M_{QL} of Chord, ST-Flooding and Ex-Flooding is 1,536, 5,232 and 14,120 ms, respectively. M_{QL} of Random walk is about 10^5 ms due to the search step limitation of 2,000. Figure 5.15 plots M_{NT} with different number of queries per vehicle. M_{NT} of HERO is much less than that of other schemes. In addition, it declines as the number of queries increases. It can be seen that, with this

Fig. 5.15 Network traffic per
query versus the number of
queries

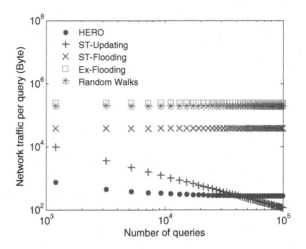

Fig. 5.16 Network traffic
rate versus the number of
queries

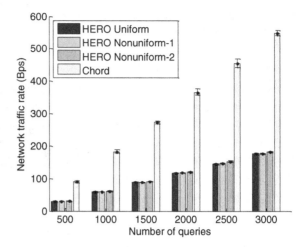

experiment setting, HERO has less query overhead than ST-Updating until the
number of queries for the same vehicle exceeds 41,400. It is very interesting to
find out that the number of queries decides whether ST-Updating or HERO is
preferable. However, we argue that it is impractical that a single vehicle would be
queried so tensely within one hour in a region with 1,000 nodes.

To further compare HERO with Chord, we conduct another experiment. We
take the same setting as the previous experiment except we distribute all the
queries in both uniform and non-uniform manners and collect the total incurred
network traffic. To non-uniformly distribute queries, we divide the whole network
into 10 areas and assign each area a different probability for a local node to
generate a query. For each probability configuration, we repeat the experiment 10
times. Figure 5.16 shows the total network traffic rate with different number of
queries. It can be seen the difference between the results of uniform distribution of
queries and non-uniform distribution of queries is very slight. We notice Chord has
incurred much more network traffic than HERO. This is because Chord takes on
average 16 hops to forward a query when the size of the logic ring is 2^{32} while
HERO guarantees to route a query within a desired number of hops (i.e., 9 hops in
this experiment). HERO takes on average 5.28 hops to forward a query in this
setting.

5.7.4 Impact of Vehicle Quantity

In this experiment, we investigate the impact of the vehicle quantity on the system
performance. We use the same network topology and protocol parameter configuration. To gain enough trace data, we take trace data from different dates and treat
a taxi at two different dates as two separate taxies. In this way, we gain 20,000

Fig. 5.17 Network traffic
rate versus individual
vehicles

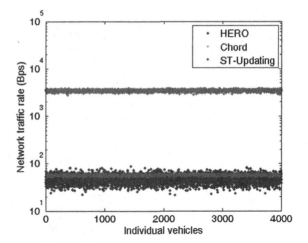

taxies traces of one hour extensive GPS data of 1,000 taxies from August 12 to
September 12 in 2006.

Figure 5.17 shows the network traffic rate with 4,000 thousand different taxies.
We can see different taxies have introduced different network traffic for location
updating. This is because different taxies have different moving patterns (for
example, a vacant taxi compared to an occupied one) and therefore the variance of
location updating cost can be large. Chord uses consistent hash to uniformly
distribute updating traffic among all local nodes, which is less influenced by different moving patterns. Beyond all the facts, we notice HERO has less average
updating traffic than Chord. This is because HERO fully leverages the inherent
locality of vehicle movements and tries to constrain updating traffic only within
nearby nodes. Moreover, we vary the total number of taxies from 500 to 3,000
with an increment of 500. Figure 5.18 plots the network traffic rate with different
number of vehicles in the system which further confirms our analysis. The average
network traffic for location updating of HERO is 47.4 Bps whereas that of Chord is
57.38 Bps. It can be seen HERO has less updating traffic than Chord and has good
scalability with the increasing number of vehicles.

5.7.5 Impact of Network Scale

To evaluate the impact of network scale, we conduct an experiment on multiple
topologies. We adopt one hour trace data of 100 vehicles, randomly generate 10^5
queries and set the real-time constraint to 500 ms. For each topology, we set r and
k according to particular numerical computation results.

In Fig. 5.19 we plot M_{QL} over different number of nodes in the system. HERO
meets the real-time constraint under different network scales. M_{QL} of ST-Flooding
increases from 5,232 ms in the 1,000-node topology up to 11,088 ms in the 3,000-

Fig. 5.18 Network traffic
rate versus the number of
vehicles

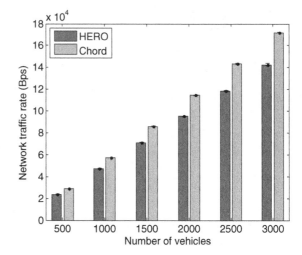

Fig. 5.19 Maximum query
latency versus the number of
local nodes

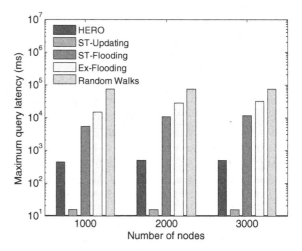

node topology. M_{QL} of Ex-Flooding also increases from 14,120 to 31,056 ms.
Random walks can always reach the maximum number of search steps. Figure 5.20 plots M_{NT} over different number of nodes. As the network scale increases,
the network traffic of HERO increases very slowly. The reason is that HERO can
constrain the updating traffic within a small region and has little influence on other
nodes which are far away from the vehicle in the network.

Fig. 5.20 Network traffic per query versus the number of nodes

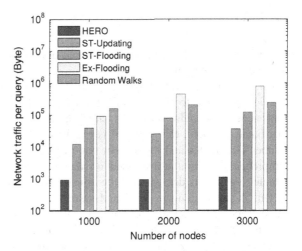

Fig. 5.21 Network traffic per query versus real-time constraint

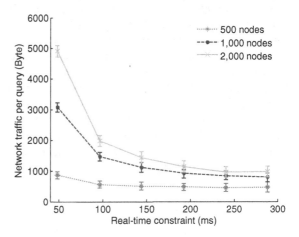

5.7.6 Effect of Real-Time Constraint

We conduct an experiment to study the relationship between the network traffic overhead per query and the real-time constraint in HERO. We use the same trace data and randomly generated queries as the experiment described in Sect. 5.7.2. We vary the real-time constraint from 50 to 500 ms. Figure 5.21 plots M_{NT} over different real-time constraints. M_{NT} first drops rapidly in the beginning and tends to increase slowly with the real-time constraint. This is because the network traffic for routing queries takes more account into the overall network traffic as the real-time constraint increases. This result is valuable for applications to select appropriate real-time constraints to satisfy their requirements while reducing the system overhead.

5.8 Summary

In this chapter, we have presented the real-time tracking protocol HERO for the metropolitan-scale intelligent transportation system. Exploiting the locality of vehicle movements in the urban area, HERO adaptively updates the locations of a vehicle according to the innovative hierarchical structure. HERO significantly reduces network traffic while still satisfying the real-time requirement. As a fully distributed protocol, this protocol is highly scalable to the number of users, the number of vehicles and the system scale as well. Prototype implementation and comprehensive simulations based on the real road network and trace data of vehicle movements demonstrate the efficacy of HERO.

This is an on-going research and system effort in tracking various vehicles in the metropolitan-scale system. Following the current work, we have a lot of more exciting yet challenging topics ahead. One of these topics is the privacy implications of tracking personal vehicles all the time. The government will guarantee to protect individual privacy by authorizing legal individuals and corporations with different privileges to access appropriate vehicles. Next, we will delve into designing better location updating schemes such that update overhead can be reduced as much as possible. Based on our realistic prototype test-bed, we will validate our design and study its performance under real complex environments. Improvements will be made based on the realistic studies before it comes to be deployed in the large-scale SG system. Moreover, it is important to ensure the security of chasers. If a malicious chaser does exist, the system may behave abnormally and the system performance would be degraded. However, this chapter focuses on the system design for real-time tracking. We will gradually incorporate security measures into the system implementation in the future.

Chapter 6
Traffic Condition Sensing Application

6.1 Introduction

One of the most important tasks in the SG project is to determine traffic condition of the road networks. The significance of this task shows in two aspects. On one hand, it provides the foundation for infrastructure construction planning as well as design optimization of public transportation systems like bus network and metro system. On the other hand, it provides the public with valuable information to plan their travels and to reduce overhead on roads.

However, it is very challenging to determine traffic condition in a metropolis like Shanghai. First, traffic condition is time-varying. Moreover, the changing of traffic condition is often unpredictable as there are so many possible factors influencing the traffic such as incidents, infrastructure construction, weather and festivals. The provided information of traffic condition would be no use if the system needs a long period of time to make the estimation. Second, it is hard to determine traffic condition of the whole road networks. In Shanghai, there are more than 22,413 intersections which connect about 33,290 road segments. It is nontrivial to provide accurate traffic information on all of these road segments.

In industry, there are already a great number of efforts aiming to provide such valuable traffic information. One simple solution would be using radio to broadcast congestion information. In such a system, congestion can be detected by eyewitness reports from commuters or news organizations. Such reports can be very coarse-grained in terms of congestion locations and duration. Many ITSs have deployed considerable sensors such as closed-circuit cameras and vehicle loop detectors as infrastructure [100, 101]. Unfortunately, the coverage of these systems is supremely limited due to the high deployment and maintenance costs. It is practically infeasible to install traffic monitoring systems densely enough to cover the entire road networks.

Instead, we propose a systematic approach, called SEER, to traffic perception on a metropolitan scale. Our approach is made up of several components. First, we define an expressive metric to reflect the traffic condition at a given site. It is not straightforward to define a good traffic condition because there are no obvious

H. Zhu and M. Li, *Studies on Urban Vehicular Ad-hoc Networks*,
SpringerBriefs in Computer Science, DOI: 10.1007/978-1-4614-8048-8_6,
© The Author(s) 2013

criteria. We define a metric, *transit velocity*, as the maximum speed at which vehicles can safely transit the site. Intuitively, a high transit velocity within the speed limitation implies a good traffic condition. Second, we deploy a cost-effective system of taxi traffic sensors. On each taxi, a GPS receiver is installed. It periodically reports its instantaneous speed and location information to a data center. Therefore, while moving around in the city, taxies act as mobile sensors perceiving surrounding traffic condition. Third, with the availability of taxi sensory data collected throughout the city, we propose an efficient algorithm which can answer traffic condition queries at any site in the city at any time. In particular, it can even make accurate prediction of traffic in a short period.

It is difficult to determine traffic condition by directly using the sensory data. First, the taxi sensory data is erroneous. The GPS location data is often not accurate, and the error can be as large as 100 m. As a result, it is difficult to map such a sensor data back to the road map. Second, sensory data may vary from taxi to taxi significantly even they are report at the same location and at the same time. In other words, each sensory data report is associated with a certain degree of noise. Third, the data is lossy and not uniformly distributed both in time and in space. For example, there are 90 % of roads that do not have sensory data for more than 80 % of all the 1,440 min in a day. The fraction would not be less than 50 % when count the number of roads that are short for data for more than 12 h in a day. In addition, we have observed that about 80 % sensory data are collected from only about 20 % roads.

Fortunately, we have observed that there are strong correlations of traffic condition over both time and space. By using conditional entropy and mutual information, we find out that knowing the traffic condition in the past does help determine the current traffic condition. Moreover, the traffic condition evolves in a basic periodicity of 1 day. Along with the spatial dimension, we notice that the traffic condition at a site has strong correlation with the traffic condition of a limit number of other sites.

For making use of the natural spatio-temporal correlations, we employ multi-channel singular spectrum analysis (MSSA) as an integral part of our solution. MSSA is a nonparametric algorithm that can effectively eliminate noise from the real signal in a time series. In our problem, the real traffic condition is our signal, and each sensory report deviates from the real signal to a certain extent. Furthermore, it provides the facility to recover signals in face of missing data. However, there are two key questions need to be answered when using MSSA in our problem. The first is how to determine the number of dimensions of the vector space that MSSA embeds the time series into. We find the optimal parameters by setting the number of dimensions to the basic periodicity of 1 day. The second is how many channels are required for MSSA to estimate the traffic condition at a site. By the spatial correlation analysis, we identify the minimum number of channels and thus reduce the computation overhead in a great deal.

In the remainder of this chapter, we describe the characteristics of the taxi sensory data in Sect. 6.2. Section 6.3 presents the spatial and temporal correlations of traffic condition we have observed. We demonstrate utilizing MSSA to estimate traffic condition in Sect. 6.4. Section 6.5 describes the methodology to evaluate the performance of SEER and presents the results. We have some discussion in Sect. 6.6. Finally, we present concluding remarks and outline the directions for future work in Sect. 6.7.

6.2 Characteristics of Taxi Sensory Data

Before we start to determine traffic condition based on taxi sensory data, it is helpful to understand the unique characteristics of the data.

In the city setting with dense high buildings and viaducts, the GPS reports from taxies can be very erroneous. The error of reported locations can be as large as 100 m. To tell which road a taxi is actually monitoring, we need to recover each sample back on track. We deal with this problem using map-matching. To the best, we can accurately recover about 90 % of the data with the left regarded as an inevitable source of noise.

In addition, we also find that individual reports vary significantly even they are collected from the same location at the same time. Figure 6.1 shows the cumulative distribution function (CDF) of speed difference derived from reports at the same location at the same time. It can be seen that the CDF increases slowly with a relatively long tail, which implies the individual reports can vary largely. The derivation of this variance may be ascribed to individual driving behaviour. For example, a taxi may stop arbitrarily to pick up or drop passengers. In other words, each sensory data report is associated with a certain degree of noise.

Further, we consider the spatial distribution of taxi sensory data. Figure 6.2 shows the number of samples on each road in a week from Dec 15 to Dec 22, 2006. Totally, there are 42,722 road segments covered by samples of 4,450 taxi traces. It is clear to see that most of the GPS samples are scattered in the

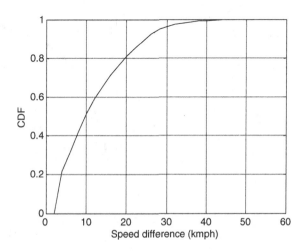

Fig. 6.1 CDF of speed difference at the same location at the same time

Fig. 6.2 Spatial distribution
of GPS samples for 1 week
from Dec 15 to Dec 22, 2006

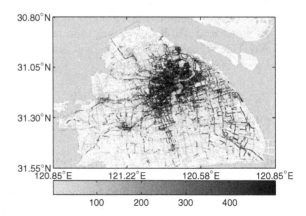

downtown area where taxies congregate more densely than in suburbs. The CDFs
of sample density on each road are shown in Fig. 6.3. The data are taken on a
weekend, on a workday and for a whole week, respectively. We observe an
obvious Pareto distribution in which the "80-20 rule" [102] stands (i.e., 20 % of
the road segments owns 80 % of the data).

Next, we concern the distribution of taxi sensory data in time. We are interested
in the probability distribution of the *inter-report times*, which refers to the time
intervals between any two consecutive reports received from a location over time.
Figure 6.4 shows the complementary cumulative distribution function (CCDF) of
inter-report times. It can be seen that the middle part of the CCDF is almost linear
in log–log scale, which indicates a power law. This means a location may frequently has no sensory data for a long time. Figure 6.5 shows the CCDFs of the
proportion of time with no sensory data in a day in different observation granularities. The time windows used to collect sensory reports are 1, 30 and 60 min,
respectively. It shows that about 90 % of roads have no samples in 80 % of the

Fig. 6.3 CDFs of sample
density at each road

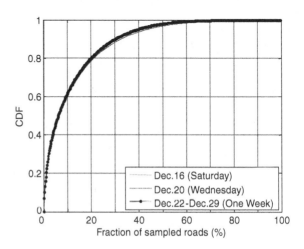

Fig. 6.4 CCDF of inter-report times

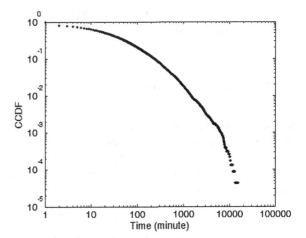

Fig. 6.5 CCDFs of the proportion of time with no sensory data

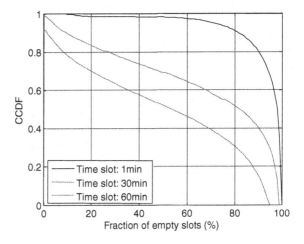

1,440 min in a day. The fraction is about 50 % when counting the number of road segments that are short of samples for 12 h in a day.

In summary, there are three crucial characteristics of the taxi sensory data with respect to using these data to determine traffic condition. First, the sensory data are erroneous in terms of large location deviation. Second, individual driving behavior introduces noise in the sensory data. Last, the distribution of the taxi sensory data is very lossy and non-uniform in time and space.

6.3 Unveiling Spatio-Temporal Correlation

According to the above study, it is not feasible to directly derive traffic condition simply from the trace data due to noise and the sparseness of sampling. In this

section, we first model the problem. Then we examine the spatio-temporal correlations of traffic condition.

6.3.1 Problem Modeling

Speaking of traffic condition, we are concerning about the transportation capability
of the road networks. It is not straightforward to define a good traffic condition
because there are no obvious criteria. We define a metric, *transit velocity*, as the
maximum speed at which vehicles can safely transit a location. Intuitively, a high
transit velocity within the speed limitation implies a good traffic condition.

Let four-tuples (*id, location, time, speed*) denote the GPS reports, where *id* is
the identifier of a taxi, *location* is the current coordinates of the taxi, *time* is the
report time and *speed* is the instantaneous speed of the taxi. We collect all the
reports from each taxi during a time window, denoted by *T*, and get the set of
sensory data, denoted by *D*. We say a road is *covered* if there is a report that is
issued from this road. Let *R* denote the set of roads that *D* has covered. With the
metric of transit velocity, we define our problem of traffic perception as: *Given the
set of sensory data D, how to determine the transit velocity at any location in the
road set R at any time in the time window T?*

There is no instant answer to this problem because of the innate characteristics
of the sensory data. Even there are sufficient reports obtained at the queried
location and time, it is hard to determine what the transit velocity is due to the
existence of noise. It can be more difficult to answer the problem when there are no
reports available.

In the following subsections, we first measure transit velocity using average
speed of reports. We then try to characterize the correlations of average speeds
over time and space.

6.3.2 Characterizing Temporal Traffic Correlations

To measure the transit velocity at location *l* at time *t*, we calculate the *average
speed* of reports which are obtained from a distance interval centered at *l* and a
time interval centered at *t*. We refer to the length of the distance interval and that
of the time interval as the calculation granularity, denoted by (Δs, Δt). With a
calculation granularity, we can establish a neighborhood of a location and divide
continuous time into separate time slots. Formally, the average speed at location
l at time *t* can be calculated as:

$$T_t(l) = \begin{cases} \sum_{i=1}^{n} v_i/n & if \quad n \geq m \\ NaN & otherwise \end{cases}, \tag{6.1}$$

where n is the total number of reports collected from the neighborhood of l in the time slot of t, v_i is the speed of the ith report and m is the minimum number of reports to calculate the average. We set m larger than one (e.g., at least three reports) to reduce the impact of individual driving behavior. If there is no sufficient reports available, i.e. $n < m$, the average speed is left blank with no value assigned.

Let us look at a simple case where we relax the spatial calculation granularity Δs. We consider average speeds on a *road segment*, which refers to the part between two neighboring intersections of a road in one direction.

For ease of computation, we further discretize the continuous average speed values into Q disjoint sub-intervals without losing generalization. We hereafter use sub-intervals to represent corresponding average speeds. For example, if the average speed value is 48 kmph and the speed values are separated by 10 kmph, we say the average speed is 4 kmph.

In order to understand how traffic condition evolves over time, we need to answer two specific questions, namely (1) how historical information is related to the current traffic condition and (2) how much historical information are related to determine the current traffic condition.

We first examine whether or not knowing the traffic condition on a road segment in the past can help us determine the current traffic condition on that road segment. We do this by computing the entropy of average speeds on each road segment and the conditional entropy of the average speed on a road segment given previous M average speeds. Let X be the random variable representing the average speeds on a road segment r. If we have observed the road segment for N time slots, the time series of average speeds can be denoted by a vector $V_r = (k_0, k_1, ..., k_{N-1})$ where $k_i \in [0, Q-1]$, $0 \leq i \leq N-1$ is the average speed in time slot i. Assume each of these Q average speeds appeared s_j times in V_r, $0 \leq j \leq Q-1$. Thus, the probability of the average speed on the road segment being j can be computed as s_j/N. Therefore, the entropy of X is:

$$H(X) = \sum_{j=0}^{Q-1} (s_j/N) \log_2 \frac{1}{s_j/N}. \tag{6.2}$$

When $M = 1$, let X' be the random variable for the immediately previous average speed on the road segment given the average speed X. X' and X have the same distribution when N is large enough. The vector V_r can be written as $W_r = \{(k_i, k_{i+1}) : 0 \leq i \leq N-2\}$. Therefore, the joint entropy of X' and X can be computed as:

$$H(X', X) = \sum_{(x', x) \in W} P(x', x) \log_2 \frac{1}{P(x', x)}, \tag{6.3}$$

where $P(X', X)$ is the number of times (x', x) appears in W_r divided by the total umber of elements in W_r. With $H(X)$ and $H(X', X)$, the conditional entropy of X given X' is:

$$H(X|X') = H(X', X) - H(X') = H(X', X) - H(X). \tag{6.4}$$

When $M = 2$, let X'' denote the random variable representing the distribution of the previous *two* average speeds given X. Similarly, the conditional entropy $H(X |X'')$ is:

$$\begin{aligned} H(X|X'') &= H(X'', X) - H(X'') \\ &= H(X'', X) - H(X', X), \end{aligned} \tag{6.5}$$

The joint entropy $H(X'', X)$ can be calculated similarly. We can continue the process and get the joint entropy when M is larger than two.

Figure 6.6 shows the CDFs of the mean entropy and the mean conditional entropy, for $M = 1$, 2, and 3, over all road segments. It can be seen that the conditional entropy when $M = 1$ is much smaller than the marginal entropy and that the conditional entropy when $M = 3$ is smaller than that when $M = 2$ which is smaller than when $M = 1$. This implies that the uncertainty about the average speed decreases when the previous average speeds on the road segment are known.

To answer the second question, we examine the correlation between the average speed in time slot t and that in time slot $t - n$ and vary n from one to a large number. Let Y_n denote the random variable for the average speed in the previous nth time slot given the average speed X. Then the condition entropy of X given Y_n is:

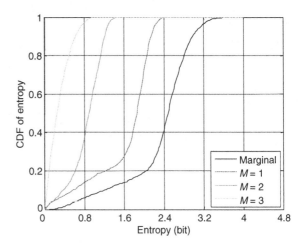

Fig. 6.6 CDFs of marginal entropy and conditional entropy of average speeds

Fig. 6.7 Conditional entropy
given average speed in each
time slot in the last week

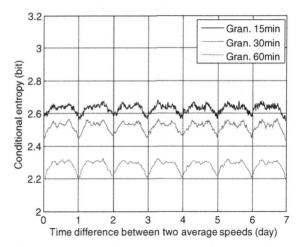

$$H(X|Y_n) = H(Y_n, X) - H(Y_n) = H(Y_n, X) - H(X). \qquad (6.6)$$

Figure 6.7 shows the conditional entropy for each time slot in previous week. The average speeds are computed with temporal granularity of 15, 30 and 60 min, respectively. In each case, we observe that the conditional entropy reaches a minimum when the value of n is times of 24 h. This means that the uncertainty about the average speed on a road segment is least when we know the average speeds at the same time on past days. Easily, we can identify a periodicity of 1 day.

6.3.3 Characterizing Spatial Traffic Correlations

In this subsection, we examine whether or not knowing the traffic condition in the neighborhood of a road segment can help us determine the traffic condition on that road segment.

We quantify the correlation between average speeds on two different road segments as follows. Recall that each road segment has a time series of average speeds. Let X_{r1} and X_{r2} denote the random variables for the average speeds on road segment $r1$ and $r2$, respectively. We can obtain the mutual information of X_{r1} and X_{r2}, $I(X_{r1}, X_{r2})$, via the joint entropy $H(X_{r1}, X_{r2})$ and the marginal entropy $H(X_{r1})$ and $H(X_{r2})$ as follows:

$$I(X_{r1}, X_{r2}) = H(X_{r1}) + H(X_{r2}) - H(X_{r1}, X_{r2}). \qquad (6.7)$$

We define the redundancy of X_{r1} and X_{r2} by

$$R(X_{r1}, X_{r2}) = \frac{I(X_{r1}, X_{r2})}{H(X_{r1}) + H(X_{r2})}. \qquad (6.8)$$

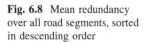

Fig. 6.8 Mean redundancy over all road segments, sorted in descending order

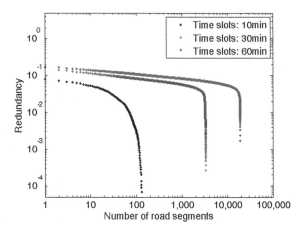

Figure 6.8 shows the mean redundancy for all road segments under three temporal calculation granularities. It can be seen that the redundancy drops dramatically at certain number of road segments at all granularities. This implies that the average speeds on a road segment are only related to a limit number of road segments. Moreover, when the granularity decreases, this number also decreases rapidly. This result is valuable when estimating average speeds leveraging spatial correlations among different road segments. We only need to consider a modest number of related road segments to estimate a road segment. Notice that these road segments do not necessarily need to be geographically near from the road segment.

In summary, we make the following observations regarding traffic perception using taxi sensory data:

- The sampling GPS data is rather lossy in terms of spatio-temporal distribution. This results frequent gaps without sufficient samples available for estimating traffic condition. We should have confidence to reconstruct the traffic condition at missing points.
- There are various sources of noise involved. This may arise from the measurement errors of GPS devices as well as from the inaccuracy of map-matching algorithms, or from the individual driving behavior. Although we can reduce this impact by aggregating multiple samples, there is impossible to remove all the noise. We should have the capability to distinguish signal from noise.
- Traffic condition has apparent spatio-temporal correlations. We find out a basic periodicity of 1 day. This periodicity has nothing to do with the calculation granularity. Moreover, we find out that a road segment has much more correlation with only a small number of road segments when the calculation granularity is small.

6.4 MSSA Based Traffic Perception

In this section, we first give an overview of our traffic perception algorithm based on *multi-channel singular spectrum analysis* (MSSA), introducing its basic rational. Next, we give a brief review to MSSA. Then we describe the iterative procedure used to estimate unknown traffic condition. Finally, we discuss the optimal configuration of the algorithm parameters.

6.4.1 Overview

Based on our above observations, we hope to extract useful information from the noisy time series of traffic condition and thus provide insight into the unknown dynamics of the underlying transportation system that generated the series.

First, we leverage the capability of MSSA to distinguish signal from noise contained in the traffic condition. MSSA takes advantage of both spatial and temporal correlations and decomposes the original time series into trends, oscillatory patterns and noise. A number of heuristic methods have been devised for signal-to-noise separation. With the trends and significant oscillatory patterns, we can reconstruct the signal.

Second, we use an iterative algorithm to deal with missing points in the time series. The algorithm iteratively produces estimates of unknown traffic condition using MSSA. Then the new estimates are used to compute a self-consistent lag-covariance matrix. The optimal window width and the minimal number of channels of MSSA are determined based on our observations on spatio-temporal correlations.

6.4.2 MSSA Review

MSSA is an ingenious application of the Karhumen–Loève expansion for random processes [103]. It provides qualitative and quantitative information about the deterministic and stochastic parts of system behavior recorded in a stationary time series even when the time series is short and noisy.

The MSSA method is data-adaptive and nonparametric based on embedding an L-channel time series with N data points $\{X_l(t) : l = 1, ..., L; t = 1, ..., N\}$ in a vector space of dimension M. A multi-channel *trajectory matrix* $\tilde{X} = (\tilde{X}_1; \tilde{X}_2; \cdots ; \tilde{X}_L)$ of X with M lagged copies can be formed by first augmenting each channel:

$$\tilde{X}_l = \begin{pmatrix} X_{l,1} & \cdots & X_{l,M} \\ X_{l,2} & \cdots & X_{l,N'+1} \\ & \cdots & \\ X_{l,N'} & \cdots & X_{l,N} \end{pmatrix}, \quad 1 \leq l \leq L \tag{6.9}$$

Thereafter, both spatial correlations between any two of L channels and temporal correlations in each channel can be obtain by computing the grand covariance matrix C_X:

$$C_X = \frac{1}{N'}\tilde{X}\tilde{X}^t = \left(C_{l,l'}\right)_{L \times L} \tag{6.10}$$

By diagonalizing the $LM \times LM$ matrix C_X, spectral information on the time series can be obtained. The eigenvectors E^k, $1 \leq k \leq LM$, of grand covariance matrix C_X are called temporal extended empirical orthogonal functions (EEOFs). Each E^k consists of L consecutive M-long segments, with its elements denoted by $E_{l,m}^k$. The eigenvalues λ_k of C_X account for the partial variance of the original time series $X_l(t)$ in the direction of E^k. Corresponding to each EEOF, space–time principal components (PCs) A^k can be computed as:

$$A_n^k(t) = \sum_{m=1}^{M} \sum_{l=1}^{L} X_{l,\,n+m-1} E_{l,m}^k, \tag{6.11}$$

where n varies from 1 to N'.

Trends, oscillatory modes and noise contained in the entire time series can then be reconstructed by using linear combinations of these PCs and EEOFs. Specifically, the kth reconstructed component (RC) at time n for channel l is:

$$R_{l,n}^k = \frac{1}{M_n} \sum_{m=L_n}^{U_n} A_{n-m+1}^k E_{l,m}^k, \tag{6.12}$$

The values of the normalization factor M_n, as well as of the lower and upper bound of summation L_n and U_n can be determined as,

$$M_n, L_n, U_n = \begin{cases} 1, 1, n, & 1 \leq n \leq M-1 \\ M, 1, M, & M \leq n \leq N' \\ N-n+1, n-N+M, M, & N'+1 \leq n \leq N \end{cases} \tag{6.13}$$

Generally, there are two main problems in using MSSA. One is how to determine the time window M (embedding dimension). The window size M should be larger than the longest periodicity that we are interested in. The other one is to determine what parts of the EEOFs are corresponding to significant oscillatory models. An oscillatory mode can be characterized by a pair of nearly equal eigenvalues and periodic eigenvectors that correspond to the same frequency.

6.4.3 Dealing with Missing Data

We adopt an iterative procedure as proposed in [104, 105] to utilize spatio-temporal correlations of traffic condition to estimate the missing points. Generally, this procedure iteratively produces estimates of missing points, which are then used to compute a self-consistent covariance matrix C_X and its EEOFs E^k. In the proposed methods, a brute-force cross-validation is required to optimize the window width M and the number of EEOFs that corresponds to significant oscillatory modes. Instead, we skip the cross-validation with confidence built on the foundation of our observations on temporal correlation. In addition, we minimize the number of channels required to reconstruct traffic condition based on our observation on spatial correlation.

Specifically, we first calculate average speeds according to a given temporal granularity. Points with no sufficient reports are regarded as missing points.

Then we set the window width of MSSA to the basic periodicity of one day. We center each channel of the original average speeds by computing the unbiased value of the mean and set the values of missing points to zero. A fraction of average speeds is left out for the purpose of validation.

Next, we start the inner-loop iteration by computing the leading EEOF E^1 and estimate the missing points using only R^1. Thus, we get a new time series with missing points estimated by R^1 and correct the mean. We then perform the MSSA algorithm again on the new series. Each estimate of the missing points is tested against the previous one until a convergence test is satisfied. Next, we perform outer-loop iterations by adding a second EEOF E^2 for estimation and repeat the inner-loop iteration. For each outer-loop iteration, we test the root-mean-square (RMS) deviation of the estimated average speeds with the reserved values. The outer-loop iterations are stopped when the minimum RMS deviations is found.

Finally, we take the parameters, K^* and M^*, that minimize the RMS deviation as the required optimum. To obtain the actual reconstruction, we repeat the inner and outer-loop iterations, using K^* and M^*, but with all available average speeds being included in the process.

6.4.4 Optimal Parameter Configuration

As described above, we establish the optimal window width of MSSA based on our previous observations on temporal correlation. This decision can greatly accelerate the search for the optimal set of MSSA parameters. Besides the window width, we also minimize the number of channels needed to estimate average speeds on a certain road segment. We do this by leverage the observation that a road segment has much more correlation with only a small number of road segments when the calculation granularity is small. Therefore, it is not necessary to build up a large grand covariance matrix C_X and enormously reduce the computation overhead.

In the next section, we validate our iterative estimation process and examine the performance on using MSSA to determine traffic condition.

6.5 Performance Evalution

6.5.1 Methodology and Metric Design

In this section, we apply the MSSA-based traffic perception algorithm to the sensory data collected from a region in the Pudong district from Dec 1 to Dec 31, 2006. Totally, there are totally 3,135 taxies involved in reporting traffic information on 235 roads.

We define the RMS deviation of two vectors of average speeds $V_1 = [v_{1,1}, v_{1,2}, ..., v_{1,n}]$ and $V_2 = [v_{2,1}, v_{2,2}, ..., v_{2,n}]$ as:

$$RMSD = \sqrt{\frac{\sum_{i=1}^{n} (v_{1,i} - v_{2,i})^2}{n}} \qquad (6.14)$$

6.5.2 Impact of Window Width and MSSA Mode Quantity

In this experiment, we investigate the impact of the time window employed in the iterative procedure on the performance of traffic perception. We set the temporal granularity to 30 min, and calculate average speeds on each road segment. We randomly choose 5 % of average speed results for validation and carry out the iterative algorithm for 20 times for window widths of 24, 48 and 60. In each run of the experiment, all of the 235 roads are employed as MSSA channels.

Figure 6.9 shows the mean RMS deviation as a function of the window width M and the number of MSSA modes. We find out that the reconstruction error drops rapidly as the number of regular oscillatory modes increases. Nevertheless, the error gradually starts to increase as the number of MSSA modes keeps growing. This can be easily understood because more PCs corresponding to noise has joined the reconstruction. We also notice that the RMS deviation reaches the globally minimum when the time window width is 48, which is 1 day in time. This result validates our suggestion of using the basic periodicity found in traffic condition as the optimum time window width in MSSA.

Fig. 6.9 RMS deviation as a function of window width M and number of MSSA modes

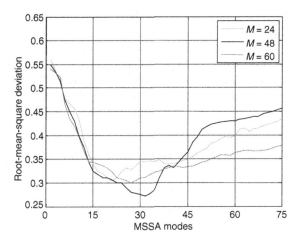

6.5.3 *Impact of Channel Quantity*

In this experiment, we examine whether we can reduce the number of channels to be used in MSSA to reconstruct traffic condition at a road segment. We randomly choose 20 road segments for traffic perception. For each road segment, we rank all road segments according to the redundancy calculated using the method mentioned in Sect. 6.4. Then we gradually add the number of road segments to be involved in the iterative algorithm. Different temporal granularities are used to calculate average speeds.

Figure 6.10 shows the RMS deviation as a function of the number of road segments involved. It can be seen that, in general, the reconstruction error deceases as the number of road segments increases. Particular, when a small temporal granularity is used, only a small number of road segments can help decrease the

Fig. 6.10 RMS deviation as a function of number of roads

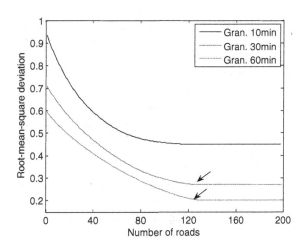

RMD error. We also notice there are slop breaks in the dash lines in Fig. 6.10 as pointed out by the arrows. The reason may be that we only choose a small region to conduct the experiment and omit some related road segments. In addition, we can also find that using the road segments after the slope breaks does not provide any more benefit but computation overhead. This result strongly agrees with our analysis on spatial correlation of traffic condition.

6.5.4 Impact of Temporal Granularity

In the above experiment, we can find out that the reconstruction error increases when the temporal granularity gets small. Reducing the granularity will cause more missing points and low signal-to-noise ratio. As the amount of mission points and noise increases, the significant PCs are "polluted" more, making it more difficult to remove the noise contributions. Even in this case, we find that the regular oscillatory modes can be determined correctly as long as the gap of missing data is not larger than any significant spatio-temporal correlations of traffic condition.

6.6 Discussion

In this chapter, the main coverage has focused on the innovation of a traffic perception system using pervasive taxi traffic sensors. On the one hand, we have put major efforts on establishing the prototype system of the taxi traffic sensors, which has been proved to be cost effective and successfully laid the foundation for our traffic perception algorithm. On the other hand, we have managed to extract traffic condition information from loss sensory data which by nature is erroneous and non-uniform. However, as a pioneering effort, the proposed approach is not perfect. Several problems need to be further investigated in future.

Due to the discrete nature of sensory data of traffic reports from taxis, it is impossible to acquire the exact sensor data at a given point. Thus, we have proposed the approximating method by using sensor reports in the neighborhood. It is worth more careful study on how much the neighborhood size could be. Note that a larger neighborhood results in a larger set of sensor reports. Meanwhile, however, a sensor report further from the given point provides data of lower quality. Given a certain scenario of traffic condition and density of sensory reports, we believe there should be an optimal neighborhood size. Nevertheless, it is not trivial to determine the optimal value.

It is apparent that we could derive better traffic condition information if more sensory data were available. However, more sensory data also imply higher investment on recruiting more taxis and conveying a larger volume of sensory data back to the information center. In addition, it would introduce higher computation cost for processing raw data and executing the algorithm. Nevertheless, it is important to study the tradeoff between the volume of sensory data and the ability of answering queries on traffic condition. Such tradeoff study will allow us to determine how much sensory data should be acquired given a certain use requirement on query quality.

6.7 Summary

In this chapter, we have presented the systematic approach to perceiving metropolitan traffic using a cost-effective system of taxi sensors. Although the raw sensory data collected by taxis are error-prone and non-uniform, our MSSA based algorithm can still effectively produce traffic condition of high quality. It also solves the lossy problem of sensory data in the sense that a given location may have a very limited number of sensor reports. As a result, this system overcomes many of the limitations for existing approaches, such as high cost and requirements on manpower. This system can be quickly implemented and serve the general public. The prototype system has been working and feeding valuable traffic condition information to the Transport Office of Shanghai.

With SEER having much space to improve, we will carry on our research in several directions. First, the problems discussed in the above subsection will be carefully investigated and constant improvements will be made accordingly. Second, we will expand our system to include a richer set of sensory data input, by employing buses and volunteer cars. Moreover, we will also incorporate the distribution information of traffic lights in the city, which may have a non-negligible impact on traffic condition.

References

1. "Federal Communications Commission. News Release, October 1999". FCC. Retrieved 2009-08-16.
2. "European Telecommunications Standards Institute. News Release, September 2008". ETSI. Retrieved 2009-08-16.
3. http://cartel.csail.mit.edu
4. Bret Hull, Vladimir Bychkovsky, Kevin Chen, Michel Goraczko, Allen Miu, Eugene Shih, Yang Zhang, Hari Balakrishnan, and Samuel Madden, "CarTel: A Distributed Mobile Sensor Computing System", in Proceedings of ACM SenSys, 2006.
5. Jakob Eriksson, Hari Balakrishnan, and Samuel Madden, "Cabernet: Vehicular Content Delivery Using WiFi", in Proceedings of 14th ACM MOBICOM, San Francisco, CA, Sep 2008.
6. "CafNet: A Carry-and-Forward Delay-Tolerant Network", MEng Thesis, MIT EECS, Feb. 2007.
7. Vladimir Bychkovsky, Bret Hull, Allen Miu, Hari Balakrishnan, and Samuel Madden, "A Measurement Study of Vehicular Internet Access Using Unplanned 802.11 Networks", in Proceedings of ACM MOBICOM, 2006.
8. http://prisms.cs.umass.edu/dome/dieselnet-buses.
9. John Burgess, Brian Gallagher, David Jensen, and Brian Neil Levine, "MaxProp: Routing for Vehicle-Based Disruption-Tolerant Networks", In Proceedings of IEEE INFOCOM, April 2006.
10. Aruna Balasubramanian, Brian Neil Levine, and Arun Venkataramani, "Replication Routing in DTNs: A Resource Allocation Approach", IEEE/ACM Transactions on Networking, 18(2):596–609, April 2010.
11. Nilanjan Banerjee, Mark D. Corner, and Brian Neil Levine, "An Energy-Efficient Architecture for DTN Throwboxes", In Proceedings of IEEE Infocom, pages 776–784, Anchorage, Alaska, May 2007.
12. Nilanjan Banerjee, Mark D. Corner, and Brian Neil Levine, "Design and Field Experimentation of an Energy-Efficient Architecture for DTN Throwboxes", IEEE/ACM Transactions on Networking, 18(2):554–567, April 2010.
13. Nilanjan Banerjee, Mark D. Corner, Don Towsley, and Brian Neil Levine, "Relays, Base Stations, and Meshes: Enhancing Mobile Networks with Infrastructure", In Proceedings of ACM Mobicom, pages 81–91, San Francisco, CA, USA, September 2008.
14. Aruna Balasubramanian, Ratul Mahajan, Arun Venkataramani, Brian Neil Levine, and John Zahorjan, "Interactive WiFi Connectivity for Moving Vehicles", In Proceedings of ACM SIGCOMM, pages 427–438, August 2008.

H. Zhu and M. Li, *Studies on Urban Vehicular Ad-hoc Networks*, 119
SpringerBriefs in Computer Science, DOI: 10.1007/978-1-4614-8048-8,
© The Author(s) 2013

15. Xiaolan Zhang, Jim Kurose, Brian Neil Levine, Don Towsley, and Honggang Zhang, "Study of a Bus-Based Disruption Tolerant Network: Mobility Modeling and Impact on Routing", In Proceedings of ACM Mobicom, pages 195–206, September 2007.

16. Fan Bai, Daniel D. Stancil and Hariharan Krishnan, "Toward Understanding Characteristics of Dedicated Short Range Communications (DSRC) From a Perspective of Vehicular Network Engineers", In Proceedings of ACM MOBICOM, 2010.

17. http://www.netlab.nec.de/Projects/fleetnet.htm

18. Jörg Widmer, Martin Mauve, Hannes Hartenstein, Holger Füßler. Position-Based Routing in Ad-Hoc Wireless Networks. In Mohammad Ilyas (ed.): The Handbook of Ad Hoc Wireless Networks, CRC Press, 2002, Boca Raton, FL, U.S.A.

19. Holger Füßler, Joerg Widmer, Michael Kaesemann, Martin Mauve, Hannes Hartenstein, "Contention-based forwarding for mobile ad hoc networks", *Ad Hoc Networks Journal*, Elsevier, pp. 351–369, Nov 2003.

20. Christian Lochert, Hannes Hartenstein, Jing Tian, Holger Füßler, Dagmar Herrmann, Martin Mauve. A Routing Strategy for Vehicular Ad Hoc Networks in City Environments. IEEE Intelligent Vehicles Symposium, June 2003, Columbus, Ohio.

21. Andreas Festag, Holger Füßler, Hannes Hartenstein, Amardeo Sarma, and Ralf Schmitz. FleetNet: Bringing Car-to-Car Communication into the Real World. In Proceedings of the 11th World Congress on ITS, Nagoya, Japan, October 2004.

22. http://www.network-on-wheels.de

23. https://www.car-2-car.org/car2car08/

24. H. Füßler, M. Torrent-Moreno, M. Transier, A. Festag, and H. Hartenstein, "Thoughts on a Protocol Architecture for Vehicular Ad-Hoc Networks," in Proceedings of WIT, Hamburg, Germany, March 2005, pp. 41–45.

25. M. Torrent-Moreno, A. Festag, and H. Hartenstein, "System Design for Information Dissemination in VANETs," in Proceedings of WIT, Hamburg, Germany, March 2006, pp. 27–33.

26. T. Kosch, C.J. Adler, S. Eichler, and M. Schroth, C. Strassberger, "The Scalability Problem of Vehicular Ad Hoc Networks and How to Solve it," IEEE Wireless Communications, vol. 13, no. 5, 2006.

27. Bookstein, and Abraham, "Informetric distributions, part I: Unified overview", Journal of the American Society for Information Science 41: 368–375, 1990.

28. A. Chaintreau, P. Hui, J. Crowcroft, C. Diot, R. Gass, and J. Scott, "Impact of Human Mobility on the Design of Opportunistic Forwarding Algorithms", in proceedings of IEEE INFOCOM, 2006.

29. T. Henderson, D. Kotz, and I. Abyzov, "The changing usage of a mature campus-wide wireless network", in ACM Mobicom, 2004.

30. P. Hui, A. Chaintreau, J. Scott, R. Gass, J. Crowcroft, and C. Diot, "Pocket switched networks and the consequences of human mobility in conference environments," in Proceedings of ACM SIGCOMM first workshop on delay tolerant networking and related topics (WDTN-05), 2005.

31. A. Bar-Noy, I. Kessler, and M. Sidi, "Mobile users: To update or not to update?", In Proceedings of IEEE INFOCOM, 1994.

32. A. E. Gamal, J. Mammen, B. Prabhakar, and D. Shah, "Throughput-delay trade-off in wireless networks", in Proceedings of IEEE INFOCOM, 2004.

33. G. Sharma, and R. Mazumdar, "Scaling Laws for Capacity and Delay in Wireless Ad Hoc Networks with Random Mobility", in proceedings of IEEE International Conference on Communication (ICC), 2004.

34. J. Broch, D. Maltz, D. Johnson, Y. Hu, and J. Jetcheva, "Multi-hop wireless ad hoc network routing protocols", in Proceedings of the ACM/IEEEMOBICOM, 1998.

35. C. Chiang and M. Gerla, "On-demand multicast in mobile wireless networks", In Proceedings of IEEE ICNP, 1998.

36. P. Johansson, T. Larsson, N. Hedman, B. Mielczarek, and M. Degermark, "Routing protocols for mobile ad-hoc networks—a comparative performance analysis", in Proceedings of ACM/IEEE MOBICOM, 1999.

37. E. Royer, P.M. Melliar-Smith, and L. Moser, "An analysis of the optimum node density for ad hoc mobile networks", in Proceedings of the IEEE International Conference on Communications (ICC), 2001.

38. R.Groenevelt, P. Nain, and G. Koole, "Message delay in MANET", in Proceedings of ACM SIGMETRICS 2004.

39. G. Sharma, and R. R. Mazumdar, "Delay and Capacity Trade-off in Wireless Ad Hoc Networks with Random Mobility", ACM/Kluwer Journal on Mobile Networks and Applications (MONET), 2004.

40. H. Cai and D.Y. Eun, "Crossing Over the Bounded Domain: From Exponential To Power-law Inter-meeting Time in MANET", in Proceedings of ACM/IEEE MOBICOM, 2007.

41. Tracy Camp, Jeff Boleng, and Vanessa Davies, "A survey of mobility models for ad hoc network research", Wireless Communications and Mobile Computing, Volume 2, Issue 5, pages 483–502, August 2002.

42. M. McNett, and G. M. Voelker, "Access and mobility of wireless PDA user", in Tech. rep., Computer Science and Engineering, UC San Diego, 2004.

43. A. Balasubramanian, B.N. Levine, and A. Venkataramani, "DTN routing as a resource allocation rroblem," in Proceeding of ACM SIGCOMM 2007, pp. 372–384, Aug. 2007.

44. Hongzi Zhu, Luoyi Fu, Guangtao Xue, Minglu Li, Yanmin Zhu and Lionel M. Ni, "Recognizing Exponential Inter-Contact Time in VANETs," in Proceedings of IEEE INFOCOM (Mini-conference), San Diego, USA, Mar. 2010.

45. H. Zhu, M. Li, L. Fu, G. Xue, Y. Zhu, and L. Ni, "Impact of Traffic Influxes: Revealing Exponential Inter-Contact Time in urban VANETs", IEEE Transactions on Distributed and Parallel Systems, vol. 22(8), pp. 1258–1266, 2010.

46. T. Spyropoulos, K. Psounis, and C. Raghavendra, "Efficient routing in intermittently connected networks: the multi-copy case," ACM/IEEE Transactions on Networking, vol. 16, no. 1, pp. 77–90, 2008.

47. Qin Lv, Pei Cao, Edith Cohen, Kai Li, and Scott Shenker, "Search and Replication in Unstructured Peer-to-Peer Networks," in Proceedings of the 16th international conference on Supercomputing, 2002.

48. Christos Gkantsidis, Milena Mihail, and Amin Saberi, "Random Walks in Peer-to-Peer Networks," in Proceedings of IEEE INFOCOM, 2004.

49. S. Jain, K. Fall, and R. Patra, "Routing in a Delay Tolerant Network," in Proceedings of ACM SIGCOMM, pp. 145–158, 2004.

50. V. Conan, J. Leguay, and T. Friedman, "Fixed Point Opportunistic Routing in Delay Tolerant Networks," IEEE Journal on Selected Areas in Communications, vol. 26, no. 5, pp. 773–782, 2008.

51. V. Erramilli, A. Chaintreau, M. Crovella, and C. Diot, "Delegation Forwarding," in Proceedings of ACM MobiHoc, 2008.

52. M. Shin, S. Hong, and I. Rhee, "DTN Routing Strategies Using Optimal Search Patterns," in Proceedings of ACM SIGCOMM Workshop Challenged Networks (CHANTS '08), 2008.

53. S. C. Nelson, M. Bakht, and R. Kravets, "Encounter-based routing in dtns," in Proceeding of IEEE INFOCOMM 2009, Rio de Janeiro, Brazil, pp. 846–854, Apr. 2009.

54. H. Zhu, S. Chang, M. Li, S. Naik, and X. Shen, "Exploiting temporal dependency for opportunistic forwarding in urban vehicular networks," in Proceeding of IEEE INFOCOMM 2011, Shanghai, China, Apr. 2011.

55. E. M. Daly, and M. Haahr, "Social network analysis for routing in disconnected delay-tolerant MANETs," in Proceeding of ACM MOBIHOC 2007, Montreal, Canada, pp. 32-40, Sep. 2007.

56. P. Hui, J. Crowcroft, and E. Yoneki, "Bubble Rap: social-based forwarding in delåy tolerant networks," in Proceeding of ACM MOBIHOC 2008, Hong Kong, China, May. 2008.

57. J. Pujol, A. Toledo, and P. Rodriguez, "Fair routing in delay tolerant networks," in Proceeding of IEEE INFOCOM 2009, Rio de Janeiro, Brazil, pp. 837–845, Apr. 2009.

58. T. Hossmann, T. Spyropoulos, and F. Legendre, "Know thy neighbor: towards optimal mapping of contacts to social graphs for dtn routing," in Proceeding of IEEE INFOCOM 2010, San Diego, USA, Mar. 2010.

59. T. Karagiannis, J. Le Boudec, and M. Vojnovi_c, "Power Law and Exponential Decay of Inter Contact Times between Mobile Devices," in Proceedings of ACM MOBICOM, pp. 183–194, 2007.

60. X. Zhang, J. Kurose, B. N. Levine, D. Towsley, and H. Zhang, "Study of a Bus-based Disruption-Tolerant Network: Mobility Modeling and Impact on Routing", in Proceedings of ACM/IEEE MOBICOM, 2007.

61. A. Lindgren, A. Doria, and O. Schelen, "Probabilistic Routing in Intermittently Connected Networks," Mobile Computing and Comm. Rev., vol. 7, no. 3, pp. 19–20, 2003.

62. Hongzi Zhu, Mianxiong Dong, Shan Chang, Yanmin Zhu, Minglu Li and Sherman Shen, "ZOOM: Scaling the Mobility for Fast Opportunistic Forwarding in Vehicular Networks," to appear in Proceedings of IEEE INFOCOM 2013.

63. R. Lu, X. Lin, and X. Shen, "SPRING: A Social-based Privacy Preserving Packet Forwarding Protocol for Vehicular Delay Tolerant Networks", in Proceedings of IEEE INFOCOM, 2010.

64. S. Milgram, "The small world problem," Psychology Today, vol. 1, no. 1, pp. 61–67, 1976.

65. H. Dubois-Ferriere, M. Grossglauser, and M. Vetterli, "Age matters: efficient route discovery in mobile ad hoc networks using encounter ages", in Proceedings of ACM MobiHoc, 2003.

66. Amaral, L. A. N., Scala, A., Barthelemy, M., and Stanley, H. E., "Classes of small-world networks," in Proceedings of the National Academy of Sciences of USA (PNAS), 97, 11149–11152 (2000).

67. M. E. J. Newman, "Modularity and community structure in networks", PNAS, 2006.

68. V. D. Blondel, J. L. Guillaune, R. Lanbiotte, and E. Lefebvre, "Fast unfolding the communities in large networks", J. STAT. MECH., 2008.

69. Q. Yuan, I. Cardei, and J. Wu, "Predict and relay: an efficient routing in disruption-tolerant networks," in Proceeding of ACM MOBIHOC 2009, New Orleans, USA, May. 2009.

70. L. C. Freeman, "A set of measures of centrality based on betweenness," Sociometry, vol. 40, no. 1, pp. 35–41, 1977.

71. M. Everett, and S. P. Borgatti, "Ego network betweenness," Social Networks, vol. 27, issue 1, pp. 31–38, 2005.

72. European Commission, "The Karen European Its Framework Architecture," http://www.frame-online.net/. 2004.

73. Department of Transportation of the United States, "The National Its Architecture Version 5.1," http://itsarch.iteris.com/itsarch/index.htm. 2005.

74. Ministry of Internal Affairs and Communications National Police Agency, and Ministry of Land, Infrastructure, and Transport of Japan, "Vehicle Information and Communication System," http://www.vics.or.jp/english/index.html. 2006.

75. Hongzi Zhu, Minglu Li, Yanmin Zhu and Lionel M. Ni, "HERO: Online Real-time Vehicle Tracking," IEEE Transactions on Parallel and Distributed Systems (TPDS), vol. 20, no. 5, pp. 740–752, May 2009.

76. Hongzi Zhu, Yanmin Zhu, Minglu Li and Lionel M. Ni, "HERO: Online Real-time Vehicle Tracking in Shanghai," in Proceedings of IEEE INFOCOM 2008, Phoenix, USA, 2008.

77. Shanghai City Comprehensive Transportation Planning Institute, http://www.scctpi.gov.cn/chn/chn.asp, 2007.

78. A. Bakker, E. Amade, G. Ballintijn, I. Kuz, P. Verkaik, I. van der Wijk, M. van Steen, and A.S. Tanenbaum, "The Globe Distribution Network," in Proceedings of USENIX Annual Conf., 2000.

79. Alminas Civilis, Christian S. Jensen, and Stardas Pakalnis, "Techniques for Efficient Road-Network-Based Tracking of Moving Objects," *IEEE Trans. Knowledge and Data Engineering*, vol. 17, pp. 698–712, 2005.

80. Dieter Pfoser, Christian S. Jensen, and Yannis Theodoridis, "Novel Approaches to the Indexing of Moving Object Trajectories," in Proceedings of Conf. Very Large Data Bases, 2000.

81. George Kollios, Dimitrios Gunopulos, Vassilis Tsotras, Alex Delis, and Marios Hadjieleftheriou, "Indexing Animated Objects Using Spatiotemporal Access Methods," *IEEE Trans. Knowledge and Data Engineering*, vol. 13, pp. 758–777, 2001.

82. Dan Lin, Christian S. Jensen, Beng Chin Ooi, and Simonas Saltenis, "Efficient Indexing of the Historical, Present, and Future Positions of Moving Objects," in Proceedings of the sixth Conf. Mobile Data Management, 2005.

83. Mindaugas Pelanis, Simonas Saltenis, and Christian S. Jensen, "Indexing the Past, Present, and Anticipated Future Positions of Moving Objects," *ACM Transactions on Database Systems*, vol. 31, pp. 255–298, 2006.

84. John F. Roddick, Max J. Egenhofer, Erik Hoel, and Dimitris Papadias, "Spatial, Temporal and Spatio-Temporal Databases—Hot Issues and Directions for Phd Research," in Proceedings of ACM SIGMOD, 2004.

85. Ben Y. Zhao, John Kubiatowicz, and Anthony D. Joseph, "Tapestry: An Infrastructure for Fault-Tolerant Wide-Area Location and Routing," Technical Report UCB/CSD-01-1141, University of California at Berkeley, 2001.

86. Ion Stoica, Robert Morris, David Karger, M. Frans Kaashoek, and Hari Balakrishnan, "Chord: A Scalable Peer-to-Peer Lookup Service for Internet Applications," in Proceedings of ACM SIGCOMM, 2001.

87. Antony Rowstron and Peter Druschel, "Pastry: Scalable, Decentralized Object Location and Routing for Large-Scale Peer-to-Peer Systems," in Proceedings of IFIP/ACM Conference Distributed Systems Platforms, 2001.

88. Sylvia Ratnasamy, Paul Francis, Mark Handley, Richard Karp, and Scott Shenker, "A Scalable Content-Addressable Network," in Proceedings of ACM SIGCOMM, 2001.

89. "The Gnutella Protocol Specification V0.6," http://rfc-gnutella.sourceforge.net. 2005.

90. Stephen Boyd, Arpita Ghosh, Balaji Prabhakar, and Davavrat Shah, "Gossip Algorithms: Design, Analysis, and Applications," in Proceedings of IEEE INFOCOM, 2005.

91. David Kempe, Alin Dobra, and Johannes Gehrke, "Gossip-Based Computation of Aggregation Information," in Proceedings of IEEE FOCS, pp. 482–491, 2003.

92. LoJack Corp., "Stolen Vehicle Recovery System," http://www.lojack.com/what/stolen-vehicle-recovery-system.cfm. 2007.

93. iPico Corp., "Test Report : Single-Lane Vehicle Identification with UHF RFID," http://www.ipico.com/site/iPico_100/pdf/WP_App_HighSpeedVehicleID.pdf. 2007.

94. "Transportation Recall Enhancement, Accountability, and Documentation (TREAD) Act," the 106th United States Congress, http://www.citizen.org/documents/TREAD%20Act.pdf. 2000.

95. Ltd Shanghai Super Electronic Technology Co., http://www.superrfid.net/english/. 2007.

96. Cisco Systems Inc., "Cisco Aironet 1240 Series 802.11a/B/G Access Point Data Sheet," http://www.cisco.com/application/pdf/en/us/guest/products/ps6521/c1650/cdccont_0900aecd8031c844.pdf. 2007.

97. Ltd Shanghai Telecom Co., http://www.shanghaitelecom.com.cn/. 2007.

98. Song Jiang, Lei Guo, and Xiaodong Zhang, "LightFlood:an Efficient Flooding Scheme for File Search in Unstructured Peer-to-Peer Systems", in Proceedings of International Conference on Parallel Processing, 2003.

99. The Network Simulator, http://www.isi.edu/nsnam/ns/. 2007.

100. B. Coifman, "Identifying the onset of congestion rapidly with existing traffic detectors", In Transportation Research, volume 37 of Part A, pages 277–291. 2003.

101 W. Lin and C. Daganzo, "A simple detection scheme for delay-inducing freeway incidents", In Transportation Research, volume 31A of Part A, pages 141–155. 1997.

102. Bookstein, and Abraham, "Informetric distributions, part I: Unified overview", Journal of the American Society for Information Science 41: 368–375, 1990.

103. K. Fukunaga, "Introduction to Statistical Pattern Recognition", Academic Press, New York, 1970.

104. Beckers, J. and Rixen, M., "EOF calculations and data filling from incomplete oceanographic data sets", J. Atmos. Ocean. Technol., 20, 1839–1856, 2003.

105. D. Kondrashov and M. Ghil, "Spatio-temporal filling of missing points in geophysical data sets", Nonlin. Processes Geophys., 13, 151–159, 2006.